WHO Technical Report Series

Research Priorities for the Environment, Agriculture and Infectious Diseases of Poverty

Technical Report of the TDR Thematic Reference Group on Environment, Agriculture and Infectious Diseases of Poverty

This report contains the collective views of an international group of experts and does not necessarily represent the decisions or the stated policy of the World Health Organization

WHO Library Cataloguing-in-Publication Data

Research priorities for the environment, agriculture and infectious diseases of poverty: technical report of the TDR Thematic Reference Group on Environment, Agriculture and Infectious Diseases of Poverty.

(Technical report series ; no. 976)

1. Communicable diseases. 2. Environment. 3. Research. 4. Climate change. 5. Agriculture.
6. Ecosystem. 7. Neglected diseases. 8. Poverty.
I.World Health Organization. II.TDR Thematic Reference Group on Environment, Agriculture and Infectious Diseases of Poverty. III.Series.

ISBN 978 92 4 120976 2   (NLM classification: WA 110)
ISSN 0512-3054

© World Health Organization 2013

All rights reserved. Publications of the World Health Organization are available on the WHO web site (www.who.int) or can be purchased from WHO Press, World Health Organization, 20 Avenue Appia, 1211 Geneva 27, Switzerland (tel.: +41 22 791 3264; fax: +41 22 791 4857; e-mail: bookorders@who.int).

Requests for permission to reproduce or translate WHO publications – whether for sale or for noncommercial distribution – should be addressed to WHO Press through the WHO web site (http://www.who.int/about/licensing/copyright_form/en/index.html).

The designations employed and the presentation of the material in this publication do not imply the expression of any opinion whatsoever on the part of the World Health Organization concerning the legal status of any country, territory, city or area or of its authorities, or concerning the delimitation of its frontiers or boundaries. Dotted lines on maps represent approximate border lines for which there may not yet be full agreement.

The mention of specific companies or of certain manufacturers' products does not imply that they are endorsed or recommended by the World Health Organization in preference to others of a similar nature that are not mentioned. Errors and omissions excepted, the names of proprietary products are distinguished by initial capital letters.

All reasonable precautions have been taken by the World Health Organization to verify the information contained in this publication. However, the published material is being distributed without warranty of any kind, either expressed or implied. The responsibility for the interpretation and use of the material lies with the reader. In no event shall the World Health Organization be liable for damages arising from its use.

This publication contains the collective views of an international group of experts and does not necessarily represent the decisions or the policies of the World Health Organization.

Printed in Italy

Contents

WHO/TDR Thematic Reference Group on Environment, Agriculture and Infectious Diseases of Poverty (TRG 4), 2008–2010	vii
Abbreviations	ix
Executive summary	xi

1. Introduction   1
    1.1 Rationale and context   1
       1.1.1 Systems-based approaches   3
       1.1.2 Recent resurgence of infectious diseases   4
       1.1.3 Emerging diseases   5
       1.1.4 Environmental and social determinants of infectious diseases   7
       1.1.5 Interdisciplinary research priorities   7
    1.2 Group membership   8
    1.3 Host country   8
    1.4 Think Tank members   8

2. Methodology and prioritization   9
    2.1 Selection of TRG members   9
    2.2 First TRG meeting   9
    2.3 Stakeholder consultation   10
    2.4 Second TRG meeting   10
    2.5 Third TRG meeting   11
    2.6 Prioritization process   11
       2.6.1 Literature review   11
       2.6.2 Principles of priority setting   12
       2.6.3 Multi-criteria decision analysis   15
          2.6.3.1 Rationale   15
          2.6.3.2 Methodology   15
          2.6.3.3 Research prioritization   16
          2.6.3.4 Criteria identification   16
    2.7 Transformation of TRG report into a WHO Technical Report   19

3. Human infectious diseases: categorization   21
    3.1 Vector-borne diseases   22
    3.2 Waterborne diseases   22
    3.3 Airborne diseases   22
    3.4 Rodent-borne diseases   23
    3.5 Soil-borne diseases   24
    3.6 Foodborne diseases   24
    3.7 Disease transmitted by body fluids   26
    3.8 Other possible classifications of human infectious diseases   26
       3.8.1 Socioeconomic status   26
       3.8.2 Vaccination status or underlying immune status   27
       3.8.3 Vaccine preventability   27
       3.8.4 Form of the infectious agent   27
       3.8.5 Zoonoses, reverse zoonoses, anthroponoses and epizoonoses   27
       3.8.6 Burden of disease over time   28

3.9	Infectious diseases of non-human species that indirectly affect human health	28
	3.9.1 Farmed mammals, birds and fish	28
	3.9.2 Birds, bats, bees and amphibians	30
	3.9.3 Infectious diseases of plants	31
3.10	Emerging infectious diseases	32
3.11	Infections and chronic diseases	34

4. Environmental and agricultural drivers of infectious diseases of poverty — 35

4.1	Forestry changes, ecological disruption and contamination	36
4.2	Dams, lakes and irrigation systems	39
4.3	Agricultural intensification	40
4.4	Climate change and infectious diseases of poverty	42
4.5	Other environmental and agricultural driving forces	42

5. Social drivers of infectious diseases of poverty — 45

5.1	Poverty	45
5.2	Population growth	46
5.3	Urbanization	46
5.4	Cultural forces and institutional change	47

6. Selected recent scientific advances, insights and successes — 49

6.1	One Health–One Medicine	49
6.2	Eco-biological mechanisms of interaction	50
	6.2.1 The opening of new 'ecological niches' for microbes	50
	6.2.2 Global trade in bushmeat and its interaction with infections	50
	6.2.3 Severe acute respiratory syndrome (SARS) and other bat-associated infections	51
6.3	Environmental quality and the burden of infectious diseases	51
6.4	Climate, seasonality, environmental change, geography and infectious diseases	52
6.5	Climate change and helminthiases (other than schistosomiasis)	53
6.6	The value of the socio-ecological perspective	54
6.7	Success stories	54

7. Hunger, nutrition, poverty and immunity — 57

7.1	Links between undernutrition and immunity	58
7.2	Undernutrition and infections: non-immunological links	58
7.3	Hunger and the first Millennium Development Goal	59
7.4	Tensions and synergies between agriculture and health	61
7.5	Agriculture and the Millennium Development Goals	62
7.6	Environment, agriculture and health: sectoral cooperation	62
7.7	Global action plan	63
7.8	Global information systems and databases	63

8. Environment, agriculture and infectious diseases of poverty: selected examples — 65

8.1	Vector-borne diseases	65
	8.1.1 Malaria	65
	8.1.2 Dengue fever	66

8.1.3	Chagas disease	67
	8.1.3.1 Biofuel plantations	68
	8.1.3.2 Amazon Countries' Initiative for Surveillance and Control of Chagas Disease	68
	8.1.3.3 Challenges for the future	70
8.2 Waterborne diseases		70
8.2.1	Schistosomiasis in Africa	70
8.2.2	Schistosomiasis in south-east and east Asia	71
	8.2.2.1 Schistosomiasis and climate change in China	71

9. Research priorities — 73

9.1 Criteria preferences and multi-criteria decision analysis (MCDA) results — 73
9.2 Relevant research priorities identified by others — 73
9.3 Priorities for policy-makers — 75

10. Conclusions — 81

Acknowledgements — 83

References — 87

Annex 1

Research priorities ranked 1-143, determined using the multi-criteria decision analysis (MCDA) methodology — 107

Appendix 1

Membership of Thematic Reference Group on Environment, Agriculture and Infectious Diseases of Poverty (TRG4) — 115

Appendix 2

Disease-specific and thematic reference groups (DRGs/TRGs) of The Think Tank for infectious diseases of poverty and host countries — 116

Appendix 3

Think Tank members — 117

Appendix 4

Distribution of the Think Tank leadership (*co-Chairs*) — 125

WHO/TDR Thematic Reference Group on Environment, Agriculture and Infectious Diseases of Poverty (TRG 4) 2008–2010

Reference Group Members

Professor C. Bradshaw, Research Institute for Climate Change & Sustainability, School of Earth & Environmental Sciences, University of Adelaide, Adelaide, Australia

Professor C.D. Butler, National Centre for Epidemiology and Population Health, The Australian National University, Canberra, Australia

Dr S. Gillespie, Director, RENEWAL, Coordinator, Agriculture and Health Research Platform, International Food Policy Research Institute (IFPRI), Geneva, Switzerland

Professor F. Guhl, Centro de Investigaciones en Microbiologia y Parrasitologia Tropical (CIMPAT), Bogota, Colombia (2008–2009)

Professor A.J. McMichael, National Centre for Epidemiology and Population Health, The Australian National University, Canberra, Australia (*Chair*)

Professor S.M. Sulaiman, Nile College, Khartoum, Sudan

Professor J.A. Trostle, Anthropology Department, Trinity College, Hartford, Connecticut, USA

Professor J. Utzinger, Department of Epidemiology and Public Health, Swiss Tropical and Public Health Institute, Basel, Switzerland

Professor B.A. Wilcox, Tropical Disease Research Laboratory, Department of Pathology, Faculty of Medicine, Khon Kaen University, Khon Kaen, Thailand

Dr A.L. Willingham III, Deputy Director, WHO/FAO Collaborating Centre for Research and Training on Neglected and Other Parasitic Zoonoses, Faculty of Life Sciences, University of Copenhagen, Frederiksberg, Denmark (2008–2009)

Dr G. J. Yang, Jiangsu Institute of Parasitic Diseases, Jiangsu, China

Professor X.N. Zhou, Director, National Institute of Parasitic Diseases, Chinese Centre for Diseases Control and Prevention, Shanghai, China (*co-Chair*)

Advisers

Professor J. Blignaut, Department of Economics, University of Pretoria, Pretoria, South Africa

Dr D. Grace, International Livestock Research Institute, Nairobi, Kenya

Career Development Fellow

Dr J.H. Huang, Zhejiang Agriculture and Forestry University, Hangzhou, China

Secretariat

Dr D. Kioy, Special Programme for Research and Training in Tropical Diseases, World Health Organization, Geneva, Switzerland

Dr A.M.J Oduola, Coordinator, Special Programme for Research and Training in Tropical Diseases, World Health Organization, Geneva, Switzerland

Dr J.U. Sommerfeld, Special Programme for Research and Training in Tropical Diseases, World Health Organization, Geneva, Switzerland

Dr A.L. Willingham, Special Programme for Research and Training in Tropical Diseases, World Health Organization, Geneva, Switzerland (2010)

Dr F. Zicker, Special Programme for Research and Training in Tropical Diseases, World Health Organization, Geneva, Switzerland

Abbreviations

AMCHA	Amazon Chagas Disease Control Initiative
BCG	Bacillus Calmette–Guérin
BSE	Bovine spongiform encephalopathy
CAFO	Concentrated animal feeding operation
CGIAR	Consultative Group on International Agricultural Research
DHF	Dengue haemorrhagic fever
DNA	Deoxyribonucleic acid
DRG	Disease Reference Group
ENSO	El Niño Southern Oscillation
FAO	Food and Agricultural Organization of the United Nations
GIS	Geographic Information Systems
IFPRI	International Food Policy Research Institute
IOM	Institute of Medicine
MCDA	Multi-criteria decision analysis
MDGs	Millennium Development Goals
NGO	Non-government organization
OIE	Office International des Epizooties
PCR	Polymerase chain reaction
SARS	Severe acute respiratory syndrome
TB	Tuberculosis
TDR	UNICEF/UNDP/World Bank/WHO Special Programme for Research and Training in Tropical Diseases
TRG	Thematic Reference Group
UNDP	United Nations Development Programme
UNEP	United Nations Environmental Programme
UNFCCC	United Nations Framework Convention on Climate Change
vCJD	variant Creutzfeldt–Jakob disease
WBLP	World Bank Loan Project
WHA	World Health Assembly
WHO	World Health Organization

Executive summary

The Thematic Reference Group on Environment, Agriculture and Infectious Diseases of Poverty (TRG 4) is part of an independent think tank of international experts, established by the Special Programme for Research and Training in Tropical Diseases (TDR) to identify key research priorities. The mandate of TRG 4 was to evaluate information on research and the challenges presented by interactions between environment, agriculture and infectious diseases of public health importance.

There is growing recognition of the fundamental relationship between human-induced changes to the environment and of the contribution of such changes to the emergence and spread of many types of infectious diseases. The scale of the problem together with the many interconnections between the various drivers of the process present enormous challenges to twenty-first century public health. Explored in this report are the benefits and limitations of a more systems-based approach to conceptualizing and investigating this problem. In this respect, a multi-criteria decision analysis approach to determining research priorities — the product of a year-long process of discussion and deliberation — is described, together with appropriate policy responses.

The dissolution of social and ecological barriers in recent decades has markedly increased opportunities for contact between humans and the animal hosts and reservoirs of pathogens. This in turn has amplified the process of microbial genetic exchange. Increases in human population and in per capita levels of consumption, facilitated by globalization, are the principal drivers of these changes. Such human domination and disruption of many of the biosphere's systems and cycles — involving urban expansion, land clearance, agricultural intensification, habitat degradation, and the industrialization of many developing countries — is without historical precedent. These complex and adverse changes to the natural world and its ecological systems are threatening the foundations of human health and survival: food yields, water availability, constraints on infectious disease outbreaks and buffers against extreme weather and environmental events.

Recent rises in extreme weather events in many regions of the world are increasingly being viewed as a manifestation of global climate change. By causing floods, surface-water pooling, anomalous rise in temperatures and displacing animal host species, such events can facilitate the emergence or spread of infectious diseases. Climate change may also already be adversely affecting crop yields and nutrient content, and hence local economies, food availability and global food prices. It is therefore likely to be increasing poverty and compromising immune function, making people more susceptible to infection.

Poverty and undernutrition are inextricably linked with many infectious and non-infectious diseases. Furthermore, acute, chronic and repeated infections lead to poor nutrition, which often 'locks in' poverty. Today's adverse environmental changes are intensifying this infectious disease risk — as also are the growing disparities in both regional and global power, wealth and policies. For at least the foreseeable future, human economic activities will disrupt, deplete or otherwise change the natural environment, climate and ecological systems. The outcomes will adversely affect many social-ecological structures, functions and systems, and have increasingly negative effects on human morale and health.

Mitigation of these outcomes requires two approaches: first, integrated and far-reaching strategies — spanning the environment, climate, agriculture, social-ecological, microbial and public-health sectors; and second, interdisciplinary research and intersectoral action. A strong partnership between science and good governance can meet these challenges. However, it will require that people modify their way of thinking and engage beyond their own specialities, since we are facing *systemic* challenges that are amplified by the increasing interconnectedness of human populations.

A much broader-based response to the evolving patterns of infectious disease risk is needed — one that entails integrative strategies and that is environmentally sustainable, socio-ecologically sensitive and adaptive to changing conditions. Development of such an approach necessitates stronger and harmonized strategic alliances between all organizations, sectors and institutions concerned with development, environment and social justice, including public health. The agenda for public health research and practice on infectious diseases can no longer be confined to itemized and vertically differentiated approaches to their prevention, control and (perhaps) eradication. Instead, it must also encompass the large-scale environmental, demographic and social changes that characterize today's world. This will require new types and levels of understanding, situation analyses, and interdisciplinary research and intersectoral actions to monitor and assess emerging trends and relationships. In this respect, the human rights community is an important ally, not only because widespread freedom of thought and movement is integral to well-being and health, but also because social and other forms of exclusion are, at core, human rights issues.

Application of the systems-based approach described in this report should result in more collaborative, integrated strategies for the prevention and control of infectious diseases, including a more ecologically aware perspective. The most important research priorities identified for infectious diseases of poverty in relation to environment and agriculture are to:

- Develop integrated preventive public health strategies for infectious diseases of poverty
- Develop and test novel intersectoral control of neglected tropical diseases
- Influence funding agencies to support inter-disciplinary approaches to infectious diseases of poverty

- Determine how to link health, veterinary and wildlife surveillance systems
- Determine which population groups are most vulnerable to climate change
- Determine the interactions between agriculture, water use and infectious diseases of poverty
- Apply systems-based research to environmentally induced transmission pathways of vector-borne diseases
- Assess the impacts of novel approaches such as community-led total sanitation on helminth infections
- Assess the impacts of water management projects on disease
- Develop and assess community-based vector-borne disease control models

1. Introduction

As part of its ten-year strategy[1] to foster "an effective global research effort on infectious diseases of poverty in which disease-endemic countries play a pivotal role", TDR established between 2008 and 2010 a global research think tank of 125 international experts to continually and systematically review evidence, assess research needs and, following periodic national and regional stakeholder consultations, set research priorities for accelerating the control of infectious diseases of poverty. Working in ten disease-specific and thematic reference groups (DRGs/TRGs), these experts are crucial contributors to TDR's stewardship mandate for the acquisition and analysis of information on infectious diseases of poverty.[2] Their work is ultimately intended to promote control-relevant research, achieve research innovation and to enhance the capacity of disease-endemic countries to resolve public health problems related to the disproportionate burden of infectious diseases among the poor.

The Thematic Reference Group on Environment, Agriculture and Infectious Diseases of Poverty (TRG 4) addresses the nature of the intersections and interactions between environment, agriculture and infectious diseases of poverty in order to identify research priorities for improved disease control.

1.1 Rationale and context

This report reviews the connections between environmental change, modern agricultural practices and the occurrence of infectious diseases — especially those of poverty — and proposes a methodology that can be used to prioritize research on such diseases. Although there is some comprehension of the underlying and growing systemic influence of today's large-scale social and environmental changes on some infectious diseases (*1*), the significance and potential future impacts of these changes are poorly understood. Nevertheless, such changes (some of which are illustrated in Figures 1 and 2) now constitute a significant influence on the working of the Earth's systems (*2*) that will have increasing consequences for patterns of occurrence of infectious diseases. Many of these changes are illustrated in this report.

A common theme of this report is bidirectional causation, effectively "trapping" complex, linked eco-social systems in stable states that are resistant to intervention. For example, poverty is associated with ill health, low education and often with poor diets, either because of undernutrition (and diarrhoea) or

---

[1] Details of TDR's strategy can be found at: http://www.who.int/tdr/publications/about-tdr/strategy/10year-strategy/en/
[2] Details of TDR's research priority reports can be found at: www.who.int/tdr/capacity/gap_analysis

intakes that have excessive calories but insufficient micronutrients. In either case, poverty impairs health; and ill health impairs the escape from poverty. Another example is provided by a recent abundant agricultural harvest in India that has far exceeded storage capacity (3). A substantial fraction of this harvest will be wasted, due to inadequate storage. Some grain that is badly stored will be contaminated by aflatoxins and other fungi, which increases the risk of cancer (see section 4.5).

This report presents the case for a more integrated approach across sectors, research disciplines and diseases (see Figures 1 and 2), taking greater account of the increasingly widespread and systemic influences on disease emergence and spread.

Figure 1
Scope of environment, agriculture, infectious diseases and society

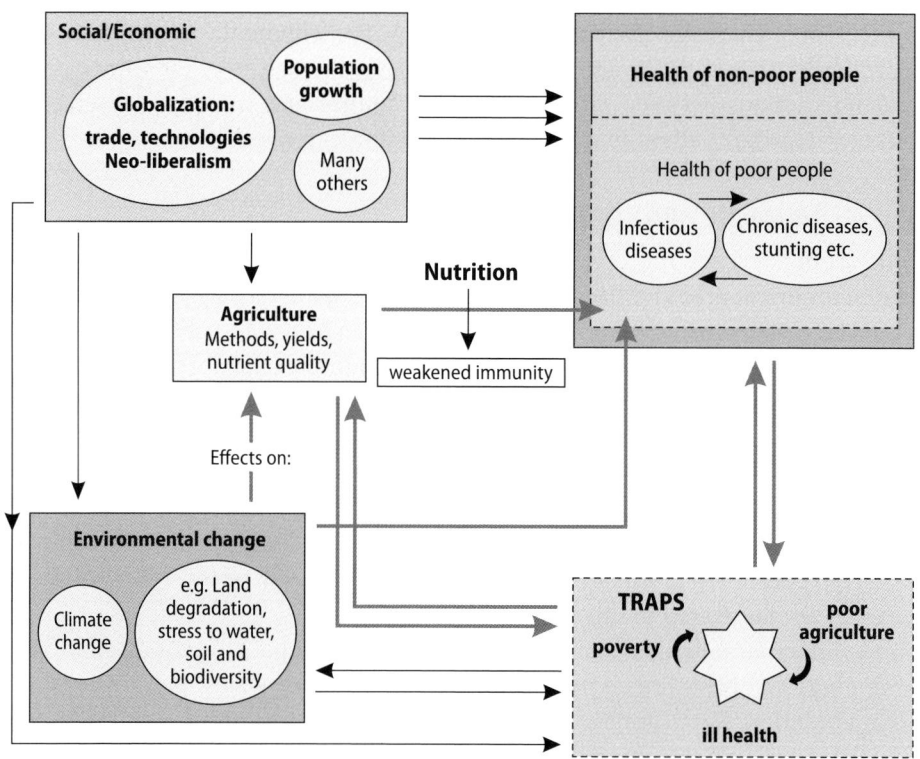

Illustrated is a complex, dynamic epidemiological landscape, reflecting the many confluent influences of larger-scale changes in the modern world that determine the global distribution of health, including the infectious diseases associated with poverty.

Figure 2
Paths between environment, agriculture and infectious diseases

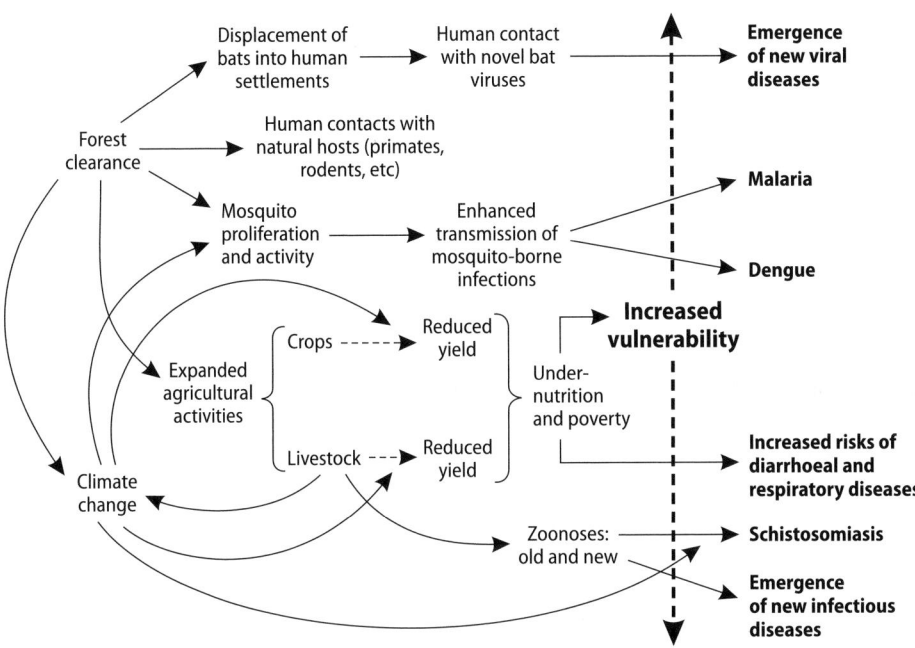

This illustrates the principle biological and ecological paths whereby agriculture and environmental change can affect the contact and transmission paths of infectious diseases, both old and new.

1.1.1 Systems-based approaches

A systems-based approach applies integrated and multidisciplinary concepts and methods to study complex, dynamic, adaptive systems such as the economy, the immune system, the ecology of malaria, and climate. Much that is relevant to infectious diseases is not reducible to simple cause-and-effect relationships. In particular, the agent–host–environment triad usually entails considerable over-simplification, especially of the environment component. The emergence and spread of an infectious disease typically reflects a complex set of relationships, most of which are non-linear (often with threshold effects) and interconnected via feedback loops. The reductionist approach to scientific analysis has had great success. However, today, the capacity of any one individual (or discipline) to understand or even to describe the workings of the world has diminished.

Broad and synthesizing understanding (albeit much less detailed) has thus given way to more narrow approaches to analysis. Reductionism necessarily leads to simplification of causal relationships. This includes the simplification of some non-linear relationships and those with feedback loops — relationships that may appear to be mathematically describable but which in reality can lead to new states that defy ready understanding, prediction, or modelling. Examples include *phase changes* (e.g. water changing to water vapour when a critical higher temperature is reached), *surprises* (e.g. a standing ovation) and *emergence* (e.g. when a pupa transforms to a butterfly). These are all *threshold* phenomena, yielding outcomes which differ *qualitatively* from their antecedents. In systems terminology, emergence can also be defined as the occurrence of complex, new phenomena from interacting, low-level mechanisms.

In practice, the most common term for this general class of threshold events is 'emergence'. A myriad of such events occur in nature. Systems-based concepts are increasingly used in ecology since they provide insights (often at least semi-quantifiable) into key pathways and processes, critical loads/stresses and hence 'tipping points' and optimal foci for intervention. The risks of Lyme disease in north-east USA have been thus described as an outcome of predator–prey relationships (wolves and deer), climatic conditions, oak-tree fruiting, deer-mouse population size, aspects of the tick's three-phase life-cycle, and human habits and habitation.

Birth, death, the onset of sickness, and good health are all examples of emergence. The term is also applied increasingly to 'emerging' diseases, i.e. those that are novel or newly recognized or which have changed host, vector or geographical range. Bovine spongiform encephalopathy (BSE) and severe acute respiratory syndrome (SARS) illustrate genuine emergence, in that their recent occurrence has no known precedence. Some other 'emerging diseases', such as HIV/AIDS and the African Ebola virus disease, have probably existed periodically in humans for millennia. However, the global scale of the HIV/AIDS epidemic is a genuine emergence.

A systems approach accepts uncertainty and imprecision, including of language. While sufficient certainty and knowledge may exist to detect robust components of the system at issue, not everything can be understood. While there is no one general 'how to' formulation of a systems-based approach, a prerequisite is an open-minded approach that can be enriched by the insights from multiple disciplines, allied with the goal of achieving social good.

1.1.2 Recent resurgence of infectious diseases

In the 1970s, improvements in living conditions and medical technology, especially vaccines and antibiotics, ushered in a global retreat of infectious diseases that was widely expected to continue (*4–9*). However, although the relative importance of certain infectious diseases in low-income settings has declined, the collective impact and burden of infection of such diseases remain high, in particular the impact of neglected tropical infectious diseases (*10*).

In many low-income countries, diseases such as dengue fever are increasing their range (*11*) and contributing to the overall burden of infectious and chronic non-communicable diseases (*12*). Tuberculosis (TB) has also undergone a resurgence, with widened transmission (*13*). Furthermore, several new or previously unrecognized infectious diseases have emerged, mostly because of changes in environmental conditions, land use, and altered food production systems. Examples include bovine spongiform encephalopathy/variant Creutzfeldt–Jakob disease (BSE/vCJD), severe acute respiratory syndrome (SARS), Nipah virus infection, various viral haemorrhagic fevers in South America, and influenza A (H1N1, swine 'flu). Large-scale outbreaks of avian influenza have also occurred, though to date with minimal transmission to humans and with little if any person-to-person transmission. The epidemiology of diseases with relationships to agriculture, including Chagas disease, Japanese encephalitis and malaria, has also changed.

Land-use change results in both benefits and harms to human health, though, from a human perspective, such changes have been overwhelmingly beneficial. Principally, this benefit has been by enabling increased food supplies, sufficient to feed the growing human population and by meeting other demands, such as for timber and other forest products. Reclamation of swamps has reduced habitat for the mosquitoes that transmit malaria and other diseases. However, land use change (including irrigation) has also accrued past, present and, increasingly, *future* health costs and risks, including those for infectious diseases. Many of these hazards are detailed in this report and range from ancient scourges such schistosomiasis and hookworm to newly recognized Henipaviruses such as Nipah virus and Hendra virus. Perhaps the most important future health risk from land use alteration is via climate change, with its likelihood of worsening nutritional status and other negative effects on global public health.

Although it has been hypothesized that the rate of emergence of infectious diseases may be accelerating, we should recognize that major cultural and demographic transitions in the past also caused regional upsurges in infectious diseases — such as during the Neolithic agricultural revolution and much later during the period of colonial expansion as a result of contacts between previously remote civilizations (each with their own pathogen pool). The Neolithic revolution refers to the integrated, systemic agricultural, technological and cultural changes which occurred independently in at least six locations in the last 12 000 years. The most important of these changes of relevance to infectious diseases are the domestication of plants and animals for food, and the consequent increase in human population density and specialization (*14*). The most notable differences in the process today are its speed, scale and global dimension, and that it is occurring in the era of modern biomedicine and public health programmes.

1.1.3 Emerging diseases

Emergence of a disease (i.e. an increase in its prevalence, geographical spread or species range) is a transient phenomenon. Following their emergence, diseases can become established and remain severe (e.g. HIV/AIDS and smallpox); slowly co-evolve with their host towards lesser virulence (*15*, *16*) (e.g. pneumonic plague and syphilis); or temporarily vanish, with occasional reappearances (e.g. African strains of Ebola virus infection).

Resistance to antibiotics and insecticides is also a form of disease emergence (see Box A). Many antibiotics occur naturally, and evolutionary processes have long shaped resistance to them. Today this evolutionary pressure is being accelerated by their widespread use, not only in humans but also in many intensively farmed animals (*17*).

A better understanding of disease emergence will foster appreciation of the need to balance environmental and commercial practices and may lessen the risk of catastrophic public health consequences (see: section 3.10).

Box A
Drug and insecticide resistance

Both drug and insecticide resistance can facilitate an increase in the incidence, spatial range and clinical severity of infectious diseases. First, the response of an infectious disease to medication will diminish if drug resistance is emerging. Second, insecticide resistance may enable a vector-borne infectious disease to spread to populations that have lost, have never had or have only partial immunity. For some diseases, such as malaria, both forms of resistance may apply. Each of these forms of resistance arises from the interaction of social, cultural, political and economic factors with the evolutionarily determined requirement of pathogens and vectors to reproduce in a changing environment. Many of these relationships are poorly understood (*18*).

Natural selection is a formidable opponent if we wish to limit or eradicate disease. While humans have inadvertently caused the extinction of many vertebrates, the much faster ability of vectors and pathogens to reproduce gives them a competitive advantage. Many pathogens also use horizontal gene transfer as a strategy to facilitate rapid and widespread evolution of drug resistance (*19*). Such transfer of genetic information obviates the otherwise slow accumulation of multiple mutations, each of which bestows only a small survival advantage, to develop resistance.

Drug and insecticide resistance can be slowed by the use of combination regimens, but in practice such strategies are vulnerable. Insecticide resistance can occur as a result of human lethargy and over-confidence, funding gaps, limited insecticide supplies and concerns about human and ecological side-effects. Donor fatigue also occurs. Integrated pest management uses a combination of biological, cultural, genetic, mechanical and chemical tactics, and has been used extensively to deal with veterinary parasites, including trypanosomes, ticks and nematodes. Although it is effective and cost-effective, the main challenge has been the high management capacity required; if this can be overcome, it may well be a model for the control of pests.

1.1.4 Environmental and social determinants of infectious diseases

Many large-scale changes in environmental, demographic and social conditions potentiate the risk of infectious diseases. Historically, these risks included the increased transmission of known diseases to naive populations. Today, this refers mainly to the emergence of new diseases with (initially) restricted transmission, the possibility of new diseases with high rates of human-to-human transmission, and the chance of known diseases reappearing on a larger scale.

The importance of socio-demographic influences has been stressed in the recent report of the Commission on the Social Determinants of Health (*20*). These factors, while currently experiencing a resurgence in recognition, have long been important (*21, 22*). In contrast, many of today's environmental influences on infectious disease occurrence have little or no historical precedent, since they result from the marked recent changes that human activities have caused to biophysical and ecological systems. These include the environmental and ecological impacts of urbanization, many aspects of globalization, patterns of land and freshwater usage, habitat destruction, intensification of livestock production and crowded livestock markets. Some of these changes cause the inadvertent introduction of invasive plants (*23, 24*), animals and pathogens ('pathogen pollution') (*25*). By enhancing antibiotic resistance, use and misuse of antibiotics add to this process of human pathogen emergence and spread. In fact, the benefits of large-scale antibiotic use in livestock as growth promoters have been exaggerated, and such use may even cause net harm (*26*).

The roles played by climate variability and change in altering infectious disease risk (as outlined above) have recently received attention (*27, 28*). However, many other environmental drivers — including dams, deforestation and habitat and biodiversity loss (*29*) — also play an important part in infectious disease emergence, and may act in tandem with climate change. For example, increases in livestock numbers, poor land management practices and the clearing of riparian vegetation to extend grazing areas increase the amounts of farm effluent, nutrients and chemicals entering rivers, lakes and coastal waters and may also lead to contamination of the environment with *Schistosoma* spp. and *Taenia* spp.

1.1.5 Interdisciplinary research priorities

Traditional, differentiated approaches to understanding and controlling the infectious diseases of poverty are no longer adequate; instead, intersectoral and interdisciplinary collaboration is vital. Conceptual frameworks are needed to shape thinking and research and assist the development and implementation of broader and more integrative policies.

A research programme that covers agriculture, the environment and infectious diseases, along with other health consequences, can increase the safety, productivity, nutritional output and income of the agricultural sector; for example, by lowering the probability of infectious disease emergence and by reducing exposure to waterborne infectious diseases, zoonoses and foodborne hazards. The adverse consequences to human health of changes in the use of land and surface water, natural habitat and climate also have to be considered.

New mechanisms are needed to foster cooperation between the organizations involved in the health and agricultural sectors. Such organizations include WHO, the Food and Agriculture Organization of the United Nations (FAO), the United Nations Environment Programme (UNEP), Office International des Epizooties (OIE), and the Consultative Group on International Agricultural Research (CGIAR). WHO brings much to such collaboration, including its political legitimacy, technical capacity, independence, direct contact with ministries of health and considerable collective experience in dealing with public health issues. Inclusion of FAO acknowledges that agriculture and health interact at the nutritional, food safety and occupational levels.

1.2 Group membership

The Thematic Reference Group on Environment, Agriculture and Infectious Diseases of Poverty (TRG 4) consisted of 12 experts in each of these three fields and the cross-cutting areas of environment, agriculture and infectious diseases of poverty (see Appendix 1), recognizing that each of these fields is a subject in its own right, and that few experts in any one have a detailed understanding of the other two. The chair and co-chair of the group were selected on the basis of their internationally recognized research and control experience in disease-endemic countries.

1.3 Host country

To ensure that the countries most affected by diseases of poverty contributed to and shared ownership of the research agenda emerging from this initiative, the reference groups were hosted by disease-endemic countries, in partnership with WHO country and regional offices (see Appendix 2).

TRG 4 was hosted in the WHO Country Office in China.

1.4 Think Tank members

The Think Tank was designed to draw on the best international expertise (see Appendix 3) and to maximize partnerships with the countries most affected by diseases of poverty. The ten reference groups making up the Think Tank include researchers and public health experts from the most affected countries, and these countries also hosted the groups. WHO country and regional offices supported both the reference groups and broad-based stakeholder consultations (see Appendix 4).

2. Methodology and prioritization

The purpose of TRG 4 was to obtain, evaluate and synthesize scientific information on global research activities and challenges in research on environment, agriculture and infectious diseases of poverty in order to provide guidance on priority research gaps and needs that should be addressed. An additional purpose was to provide independent advice and guidance on priority areas and critical research gaps as a contribution to the *Global Report for Research on Infectious Diseases of Poverty* (*30*). There are many ways to identify priorities based on expected outcomes. TRG 4 used a process of expert opinion solicitation, literature review, stakeholder consultation and eventual analysis based on multiple criteria.

2.1 Selection of TRG members

Potential members were identified from research institutions, international organizations, health, medical, nutrition and agricultural organizations, and governmental and inter-governmental organizations worldwide and evaluated by a panel of WHO internal and external experts. Particular attention was paid to the geographical distribution, to ensure disease-endemic country and regional input as well as technical input, and gender balance of the membership. The final list of members was formally appointed by the Director of TDR for an initial period of 2 years. All members were obliged to declare any conflict of interest and confidentiality.

2.2 First TRG meeting

The first meeting of the TRG 4 was held in October 2008 in Beijing, China, where its long-term operations, concept and overall scope were discussed. Members were asked to prepare presentations on research topics in their areas of expertise regarding the complex interplay of environment, agriculture and infectious diseases in order to provide a baseline for research and help identify key research gaps in order to further define the scope of the work to be pursued by the group. The TRG4 terms of reference were as follows:

- To obtain, evaluate and synthesize scientific information to enhance understanding about the linkages and synergies between environment, agriculture and infectious diseases.
- To specifically focus on infectious diseases of poverty including, but not limited to, vector-borne, soil-transmitted, waterborne, foodborne, zoonotic diseases and emerging infectious diseases threatening human health, at different spatial and temporal scales, with a particular emphasis on determinants of vulnerability such as gender, poverty and inequality.

- To evaluate the relevance of this knowledge to control, with special reference to infectious diseases of poverty.
- To provide independent advice and guidance on priority areas and research gaps.

2.3 Stakeholder consultation

Periodic regional and national stakeholders' consultations were an essential part of the Think Tank analytical process, enabling validation, endorsement and uptake of final research priorities, ensuring that the group's work was authoritative, scientifically credible, and relevant for policy. The TRG 4 research prioritization process was strengthened through several opportunities for interaction with and feedback from stakeholders that were facilitated by the WHO Representative Office in China and the WHO Regional Office for the Western Pacific (WPRO).

A stakeholder consultation involving Chinese national authorities (e.g. Chinese Ministry of Health, the Chinese Center for Disease Control and Prevention, the National Institute of Parasitic Diseases, Chinese Academy of Agricultural Sciences, China Council for International Cooperation on Environment and Development) and regional representatives of international organizations (e.g. FAO, UNEP, University of Ghana, Royal Norwegian Embassy) was held in conjunction with the first TRG 4 meeting in Beijing in October 2008. The main objectives of the consultation were to provide an overview of the scope and work plan of the TRG; to present current issues and public health challenges in environment, agriculture and infectious diseases in China; and to provide an opportunity for stakeholders to contribute to the framework of the TRG's planned activities. Stakeholders acknowledged the complexity of the inter-linkages between environment, agriculture and infectious diseases. To address this complexity, they stressed that the TRG 4 should take an innovative analytical approach, specifically the needs for interdisciplinary, inter-sectoral frameworks and a prospective, systemic and ecological approach to framing the issues. The stakeholders emphasized that TRG 4 analysis should be expressed in practical terms to policy-makers in order to lead to action.

2.4 Second TRG meeting

The second meeting and stakeholder consultation of the TRG 4, held in October 2009 in Shanghai, was entitled "Innovative Ecosystem-based Intervention for Infectious Diseases Control". It brought together members of the TRG 4 with public health practitioners and researchers in China who had had working experience with innovative ecosystem-management interventions for the control of selected infectious diseases. The consultation reviewed the first annual report of the reference group and focused on practical solutions for the control of infectious diseases that are induced by environmental and/or agricultural changes, with specific case studies being presented on a range of parasitic diseases and viral infections, including schistosomiasis, dengue fever, avian influenza, echinococcosis and fish-borne trematode infections. The stakeholders underlined the need for further research on the control of environmentally-induced infectious diseases.

2.5 Third TRG meeting

The third meeting of TRG 4 took place in October 2010, starting with a stakeholder consultation involving local public health practitioners in Chongqing city (China's fourth largest municipality/directly administered city), continuing with the TRG 4 deliberating on prioritizing the specific research needs, followed by a field visit to schistomiasis-endemic areas in Jingzhou, Hubei Province. Consensus was reached on the process for final prioritization of the research needs identified. The process was completed through electronic workspace and e-mail exchange in the weeks immediately following the meeting.

2.6 Prioritization process

2.6.1 Literature review

An overview of the available literature was obtained by soliciting expert opinion, starting with the TRG 4 members, and supplementing this by a literature search using numerous relevant keywords, embracing both general and specific topics. General terms included "agriculture", "environment", "emerging infectious diseases", "epizootics" and "infectious diseases". Many search terms linked specific infectious diseases (e.g. "Chagas disease", "dengue fever", "geohelminths", "hantavirus", "Hendra", "hookworm", "influenza", "leishmaniasis", "leptospirosis", "Lyme disease", "malaria", "nematodes", "Nipah virus", "plague", "rabies", "schistosomiasis", "*Taenia solium*", "tapeworm", "tuberculosis") with specific environmental and agricultural factors (e.g. "agricultural intensification", "aflatoxin", "aquaculture", "bats", "biodiversity", "biofuels", "bushmeat", "climate change", "deforestation", "ecology", "extreme weather events", "flooding", "irrigation" and "palm oil"). Other links were also explored, such as between some of the above terms and "complex systems", "demography", "education", "emergence", "food security", "governance", "immunity", "nutrition", "One Health", "population growth", "poverty", "systems thinking", "undernutrition" and "vectors". The literature in this area is too large to permit a systematic review, and extends far beyond that normally considered as biomedical or health related. Google Scholar was considered the most efficient tool to search these diverse literatures.

Search terms were restricted to English, though some relevant literature published in other languages was identified. Expert judgement was used to examine more closely a subset of identified literature which was still too large for all of it to be cited or discussed here. Several additional references were suggested by the reviewers. All this material was used to analyse and prepare a succinct summary of the important current issues. The key sources were peer-reviewed journal articles, supplemented by a few reports and books.

The nature of the literature necessitated a reductionist search strategy, while summarizing it required a classificatory approach. To this end, we classified diseases mainly by their main route of transmission. This approach permits readers to focus on their specialist areas, without requiring them to read in detail material that is less relevant to them. Such an approach has necessitated some repetition, but without duplication of detail. For example, while Nipah virus is mentioned in several places, the details are not repeated, but cross-referenced.

2.6.2 Principles of priority setting

Research proposals and priorities often evolve through a process of assembling and critiquing evidence, constantly searching for fresh insights by examining and testing new hypotheses. Because research funds and human expertise are limited, prioritization is essential and several criteria are relevant to this (see Figures 3–5).

Ideally, research priorities relevant to the themes covered by this report should address problems of high disease burden that are amenable to research. Such problems are, however, rare; examples in the second-most favourable category (upper right in Figure 3) are the search for high-impact vaccines, such as malaria and HIV/AIDS. Discovery of an effective vaccine for HIV/AIDS or malaria or of an improved tuberculosis (TB) vaccine would be of high benefit, but is far from easy, as evidenced by the decades of effort expended towards these goals, with only limited success. In Figure 3 such tasks therefore appear in the outer circle of the upper-right quadrant. Opportunities in the lower-right quadrant (potential for high disease-burden benefit, easy to do) are also extremely rare. However, over time, research projects can shift between quadrants, not only as disease burden changes, but also as scientific knowledge advances. One example from the lower-left quadrant is the development of an equine vaccine for Hendra virus infection — whose burden of disease is comparatively low — but which presented few technical difficulties. A further example is the relatively easy development of dipstick tests for malaria (*31*), a disease whose burden is high. In the upper-left quadrant problems appear extremely difficult, yet their solution promises little public health benefit. Research that fits these conditions is thus the least rational on which to focus effort.

Research priorities can be distorted by institutional barriers, prevailing policy agendas, external markets and by linkage to commercial opportunities (*32*). For example, research on food and nutrition in low-income countries is currently largely focused on obesity and over-nutrition, rather than on undernutrition and its associated synergies with infection. Also commercial forces channel researchers towards costly, less effective interventions to address obesity (*33*).

Methodology and prioritization

Figure 3
Research prioritization criterion 1: Global Burden of Disease

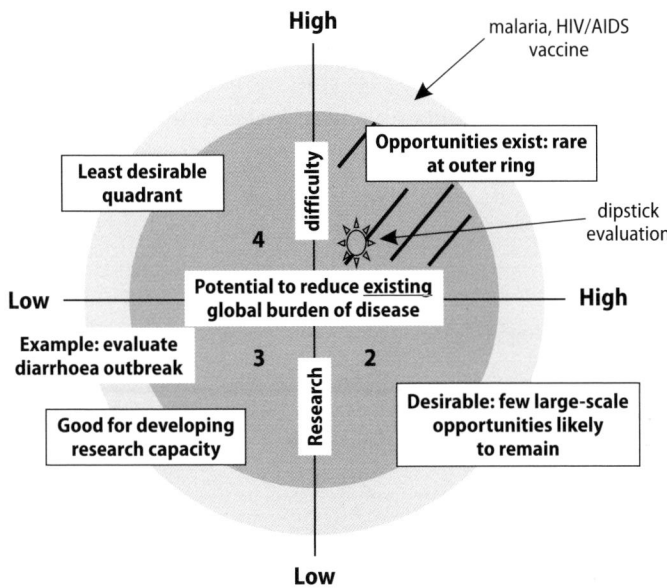

Figure 4
Research prioritization criterion 2: Potential disease burden

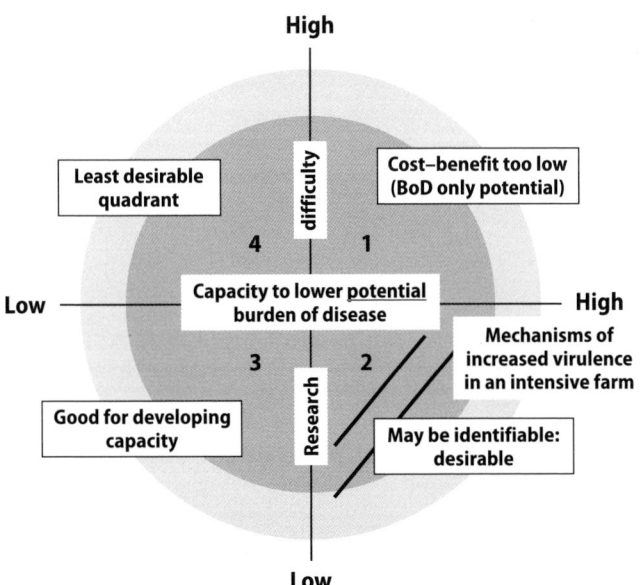

Opportunities in the second quadrant should be explored first. Again, a focus in the upper-left quadrant is least desirable. BoD = burden of disease.

The *existing* burden of disease and amenability to research are not the only criteria. Also important is the *potential* future burden of disease — especially since it presents opportunities for pre-emptive action (see Figure 4). An example of this is the investigation of mechanisms of possible enhanced pathogen virulence in intensively farmed animals (*17, 34*). This lies in the second quadrant because insights gained by research appear likely to lower a new and significant burden of disease, without posing formidable research difficulties. Raising the funds and justifying the effort to tackle even more difficult challenges is unlikely to be possible until such disease burden has moved from theoretical (potential) to actual.

Figure 5
Research prioritization: Integrated criteria

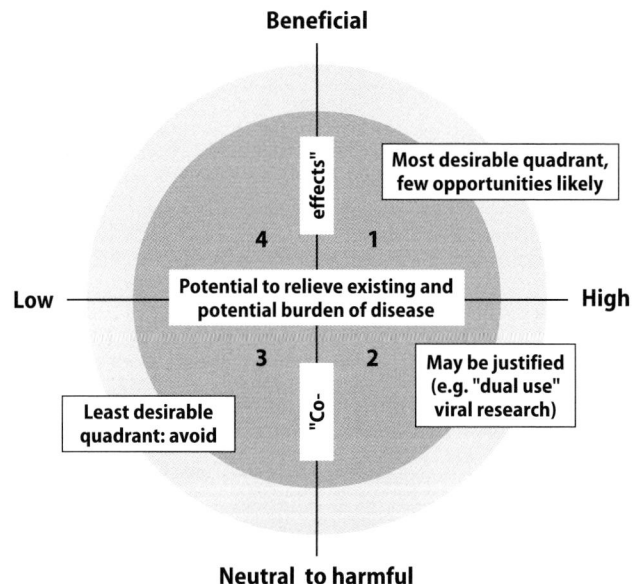

Many research activities have co-effects. These can be beneficial (e.g. empowering and informing affected societies) or harmful (e.g. exploitative).

Figure 5 shows a third important research priority criterion — the potential for gaining the "co-effect" of fundamental biological, ecological or social insights. For example, research into barriers that impede the reduction of a specific disease burden may identify factors (e.g. involving community participation) which may also apply to other health problems, a co-benefit (*35*), although some co-effects may be harmful (e.g. exploitative). Action research, which empowers communities in addition to discovering new evidence, provides another example of a research co-benefit (*36*). Some researchers argue that work that may have potential harm is justified if the benefit to reduce disease burden is very high.

2.6.3 Multi-criteria decision analysis

2.6.3.1 Rationale

Application of a systems approach to health decision-making makes the process more realistic, recognizes that human health is inter-dependent with that of animals and of the ecosystem, and views health in the broader social and environmental context. Numerous simpler approaches have been developed to assist health decision-makers, including burden of disease analysis, cost-effectiveness analysis and stakeholder consultation (*37*). However, such approaches typically rely on limited criteria, do not adequately address trade-offs between objectives that have different importance (*38*) and are not oriented towards anticipating 'surprise' outcomes.

Multi-criteria decision analysis (MCDA), which is widely used to analyse similar multifaceted problems in other disciplines, permits integration of several streams of information and helps decision-makers gain insight into the values that underlie choices and hence make more transparent and rational decisions (*39*). MCDA has also been recommended for guiding decisions about resource allocation in health (*40*).

2.6.3.2 Methodology

Of the numerous available methods for MCDA, the TRG 4 used one based on multi-attribute utility theory. This assumes that all attribute utility functions are linear, so that the total utility function (U) is a simple weighted sum of the attribute measures — a reasonable approximation when the range of attributes is relatively narrow. The method had eight main steps.

1. Specify the objective of the decision-making exercise
2. Identify an expert group
3. Identify the alternatives to be appraised
4. Identify criteria important to the decision
5. Consider the criteria
6. Score the alternative options
7. Calculate dominance scores
8. Examine the results

The objective was for a group of experts to identify the priority questions related to environment, agriculture and infectious disease of poverty to guide research over the next decade. A large set of possible research questions was generated as described below.

2.6.3.3 Research prioritization

Several priorities were identified through expert-facilitated discussion at the second TRG 4 meeting held in October 2009. Subsequently, TRG 4 members were sent a summary of the priorities (in general and in four specific areas), which they were asked to consider and evaluate before attending the third TRG 4 meeting in October 2010. All the research options were reviewed at this latter meeting and further suggestions offered. The list of options was then edited and the format standardized, resulting in 143 possible research options related to environment, agriculture and infectious diseases of poverty. The 'word cloud' in Figure 6 gives a snapshot of these by representing diagrammatically the terms used and their relative frequency.

Figure 6
Word cloud

This computer-generated diagram gives greater prominence to words that appeared more frequently in the list of identified research priorities.

2.6.3.4 Criteria identification

During the TRG 4 Workshop in October 2010, a subgroup of four experts identified fifteen measures of performance (criteria) against which the alternative research priorities could be classified and scored. These criteria had unambiguous definitions (see Table 1). They were then ranked blindly by the larger group of TRG 4 members and advisers (in total 12 experts), and the four criteria with the highest average score were retained for the MCDA. The experts were also encouraged to suggest any additional criteria, but none emerged. The raw and standardized scores for the twelve most highly ranked research criteria are shown in Table 2, together with the raw scores classified by the disciplinary background of the expert concerned (i.e. health or non-health).

Table 1
Definitions of the fifteen criteria used to assess research priorities

Criterion	Definition
Inter-disciplinarity	Involves three or more disciplines working continuously and interactively (i.e. not combining only at the end)
Impact on reduction of disease burden	Effectively targets diseases with high impact on human populations
Potential for 'other benefits'	Benefits other sectors (e.g. livestock, trade, tourism, income generation, conservation etc.)
Financial sustainability	Reduces need for recurrent expenditure
Equity	Provides preferential benefits to poor and/or socially excluded groups, e.g. women and children, minorities
Value for money	Relatively large potential benefits for relatively small research costs
Innovation	Novel concept, methodology, and/or technology (including appropriate technology)
Feasibility/practicality	Achievable, credible, testable, replicable results
Preventing disease with high potential burden	Potential aversion of high impact, low probability events (e.g. HIV)
Capacity-building potential	Improving knowledge and skill among service providers, policy-makers, communities (and students)
Systems framework	Contribution to development of a systems framework which addresses health holistically (non-reductionist)
Community focus	Research attends to, engages, empowers and/or delivers benefits to communities involved
Multi-level	Research focuses on individuals, households, populations and ecosystems and/or end-user, service provider, researcher and decision-makers
Reflects lessons learned	Opportunity for learning from and building on past successes and failures
Potential for policy impact	Policy relevance and proactive involvement of/influence on policy-makers (including MDG and other targets)
	Any criterion you think is important for making decisions about priorites and not included in the above?

Table 2
Raw and standardized scores for the twelve most highly ranked research criteria used for assessing research priorities at the interface of environment, agriculture and infectious disease of poverty[a]

	Score by experts			
Criterion	Raw	Standardized	Non-health	Health
Inter-disciplinarity	*87*	*86*	*88*	*86*
Feasibility/practicality	*78*	*70*	*87*	*71*
Impact on reduction of disease burden	*75*	*70*	*70*	*78*
Potential for policy impact	*74*	*67*	*71*	*76*
Multi-level	66	54	71	62
Preventing disease with high potential burden	65	47	57	71
Innovativeness	64	52	83	51
Value for money	64	50	72	59
Systems framework	63	51	63	62
Capacity-building	61	49	55	65
Potential for other benefits	61	48	54	65
Equity	56	45	50	60

[a] Scores shown in descending order (*n* = 12). The top four are shown in italics.

Since all the scores[3] were qualitative, we used a simple linear additive evaluation method to calculate a final score (the score for each criterion was multiplied by the weight of that criterion and all the weighted scores were summed). MCDA is an aid to and not a substitute for decision-making. The final step involved examining the overall dominance scores, which indicate the value of one alternative over another, and can be used to help assign priorities.

---

[3] The scores assigned by the experts to the four highest-ranking criteria were standardized across experts by subtracting the minimum score assigned and dividing the result by the range.
Standardized score $_i$ = (score $_i$ – minimum score) / (maximum score – minimum score)
The experts scored each of the 143 research options using these four highest-ranking criteria and the scores obtained were standardized by dividing by the highest score possible.
Standardized score $_j$ = score $_j$ / maximum score

Further insight into the robustness of the results can be obtained by carrying out a sensitivity analysis.

The results of the prioritization process through MCDA are presented in chapter 9.

2.7 Transformation of the TRG report into a WHO Technical Report

The process of finalization of the report and its transformation into a publication in the WHO Technical Report Series was carried out through electronic communication between the Chair, co-Chair, TRG members and the WHO Secretariat. The WHO Secretariat undertook the organization of external and internal reviews of the report, and comments on structure and content were addressed in the final version of the report.

This report was assessed by the WHO Guidelines Review Committee, which recommended that it be published as an Expert Report.

3. Human infectious diseases: categorization

We used a simple classification of diseases that should serve as a useful starting point, especially for readers who lack specialist knowledge of the infectious diseases of poverty.

Infectious diseases can be viewed ecologically as a host–parasite–reservoir relationship. Collectively, viruses, bacteria and protozoans have been termed 'microparasites' (*41*), in contrast to larger 'macroparasites' such as carnivores. Many organisms coexist peacefully within their hosts, and many symbiotes are beneficial, even essential to their host's survival.[4] This tolerance by hosts results from a long co-evolutionary experience that has selected against the most highly pathogenic genotypes and at the same time selected hosts that had greater tolerance for these microparasites (*16, 42*).

Over 1400 infectious disease pathogens have been identified in humans (*43*) and more will certainly emerge. These pathogens are classified here according to their dominant medium of transmission — arthropod vectors, water, food, air, soil, rodents and body fluids — although there are some overlaps and possible ambiguities (see Figure 7). This simplified figure shows the complexity but highlights the key pathways. There are two main ways in which the agriculture–environment 'complex' alters the burden of infectious diseases of poverty: either via direct exposure to new pathogens or via food yields and altered nutrition, which either enhances or lowers human capacity, including immunologically.

Figure 7
Links between the agriculture–environment 'complex' and health

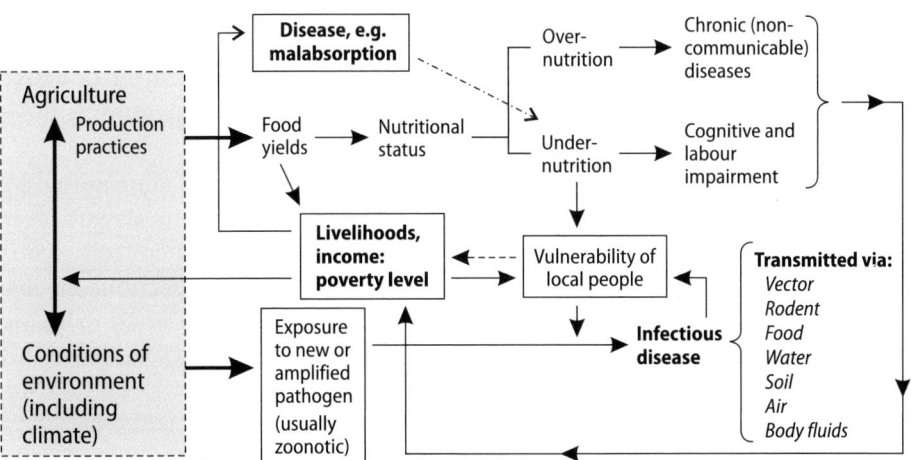

---

[4] All forms of life rely on other forms, if not for nutrition then at least for community.

3.1 Vector-borne diseases

The vectors that transmit human pathogens are predominantly arthropods, such as mosquitoes, sandflies, ticks, black flies, fleas and triatomines. Vector-borne diseases include Chagas disease, filariasis, leishmaniasis, malaria and many viral infections (e.g. Chikungunya virus disease, dengue fever and yellow fever). Onchocerciasis (river blindness) is a vector-borne disease that can also be considered to be waterborne. Transmission of vector-borne diseases to humans occurs because the life-cycle of the pathogenic organism involves an arthropod vector and a human, although other animal species (e.g. cattle) may act as principal hosts. Vectors transmit pathogens by biting or defecating on hosts, whose blood they need as sustenance.

Mammals such as dogs, bats and many rodents may also be considered to be vectors of human disease, such as rabies, diseases caused by other genotypes of lyssavirus and of Ebola virus disease. Rodent-borne diseases are briefly described later in this chapter, while several bat-associated diseases are described elsewhere in the report (e.g. sections 3.9.2, 6.2.3 and Box E). Vector-borne diseases are also important in plants (*44*, *45*).

3.2 Waterborne diseases

This category refers to diseases caused by parasites whose life-cycle involves water (*46*). Examples include dracunculiasis (guinea worm disease) and various trematode infections, including schistosomiasis caused by the six major forms of *Schistosoma* spp. that infect humans, as well as other intestinal, liver and lung fluke infections (e.g. those involving *Opisthorchis viverrini*, *Opisthorchis felineus*, *Clonorchis sinensis*, *Fasciola hepatica* and *Paragonimus* spp.). Schistosomiasis is usually classed as waterborne, since freshwater snails act as intermediate hosts for the infective agent. Unlike vectors, the snails do not participate in active transmission of the infective agent to humans. Instead, the free-swimming fluke leaves the snail and penetrates the skin of the host.

Some classifications of waterborne disease include those transmitted by water-dependent mosquito vectors, such as filariasis, malaria and dengue fever (*46*). Leptospirosis, too, could be classified in several categories: water-, soil-, food- and rodent-borne (*47*, *48*). Many other important waterborne infections involve neither a vector nor an arthropod host but are transmitted directly by drinking contaminated water; examples include cholera, cryptosporidiosis, giardiasis and many viral diseases.

3.3 Airborne diseases

Airborne diseases of humans involve transmission of the pathogenic agent (generally a virus or bacterium) from the respiratory tract of an infected animal or person via the air. Often, however, transmission occurs through contact with secretions that have fallen from the air onto surfaces, underscoring the importance of frequent hand washing in avoiding infection (*49*).

Examples of such diseases include influenza, pneumonic plague, TB and hantavirus pulmonary syndrome. Some diseases, including leprosy and TB, can be transmitted in several ways. Two of the most important infectious diseases of poverty are airborne: pneumonic plague (*50*) and viral influenza. The epidemiology of each is influenced by environmental factors, while that of some forms of influenza also have agricultural determinants.

3.4 Rodent-borne diseases

Rodent-associated diseases include plague, hantavirus pulmonary syndrome, leptospirosis, Lyme disease and tick-borne encephalitis. Arenaviruses are also associated mainly with rodents, transmission occurring principally through contact with their excreta and secreta (*51*) following human incursion into environments — including the clearing of forests to plant crops — that provide rodent habitat.

Leptospirosis ('field fever') is particularly associated with agriculture, livestock contact and flooding (*52*), but its incidence is also sensitive to more distal ecological and social changes. For example, in Sri Lanka, the recent increase in the incidence of leptospirosis may be due in part to a sustained campaign to reduce rabies by culling the population of wild dogs following the 2004 tsunami. The resultant reduced competition from dogs for waste food is thought to have increased rat numbers, while at the same time may also have reduced rat predation (*53*). Urbanization and climate change may also be enhancing leptospirosis epidemics, especially in low-lying tropical cities at risk of flooding (*54*).

The two major epidemics of plague in Europe (the Justinian plague, which started in 541, and the Black Death, which started in 1347) were both associated with high populations of *Rattus rattus*. This species is thought to have migrated to Europe from south-east Asia in antiquity, through a variety of human activities, including shipping, the distribution of grain in the Roman Empire, and the increased accumulation of waste matter as parts of the Empire descended into chaos (*55*). Climatic and nutritional factors may have also played a role in both pandemics (*56–58*).

Several rodent-borne diseases are associated with plant ecology. For example, the likelihood of contracting Lyme disease is determined in part by the abundance of acorns. In north-east India, the periodic flowering and masting of the bamboo species *Melocanna baccifera* leads to explosions in rat populations. This precipitates famine (as the rats devour crops) and increases the incidence of fevers, some of which are probably infections of rodent origin (*59*).

In Belgium, climate change has been linked to increased availability of the seeds of deciduous trees, such as acorns from oaks and nuts from beeches (mast), staples of the bank vole (*Myodes glareolus*). During the resulting 'mice years', bank vole populations can grow as much as tenfold. The trend and size of these events is increasing and has been linked with a higher incidence of a mild syndrome of haemorrhagic fever and renal failure caused by Puumala virus, a type of hantavirus (*60, 61*). In Scandinavia, too, warmer winters are thought to have promoted contact between voles and humans in barns, leading to higher levels of human disease (*62*).

3.5 Soil-borne diseases

Soil-transmitted helminthiases include infections with hookworm (*Ancylostoma duodenale* and *Necator americanus*), roundworm (*Ascaris lumbricoides*) and whipworm (*Trichuris trichiura*). Hookworm can be transmitted directly through the skin (often bare feet), but other nematodes require a faecal–oral route; lack of hand washing is an important risk factor. Helminthiases have a very high burden of disease, collectively exceeding that of all other parasites apart from malaria (*10*), and reduce cognitive development (*63*). The epidemiology of helminthiases is also likely to be altered by climate change (*64*), as may also be that of leptospirosis and melioidosis (*54, 65*).

3.6 Foodborne diseases

Foodborne disease takes many forms and infection can occur anywhere between 'farm and fork' (*66*). Contamination of food may thus occur in the field from pesticide sprays, night soil or poor hygiene, or from coprophagic animal behaviour, which can lead to inadvertent ingestion of the eggs of parasites such as the pork tapeworm, *Taenia solium*.

If not cooked properly and handled with great care, any such infected meat can cause human taeniasis, where the adult tapeworm develops in the gut. Humans may also become intermediate hosts if they ingest eggs or oncospheres, which encyst in body tissues, including muscles, eyes and the brain. A recent WHO-commissioned systematic review of studies reporting the frequency of neurocysticercosis estimated that it occurs among 29% of people with epilepsy (*67*). This is a particularly common cause of epilepsy in poor pig-keeping communities (see Box B).

In 2010 in the United States of America (USA), half a billion eggs were recalled following an outbreak of salmonellosis. Several forms of feed used in the intensive farms that produced these eggs were found to be contaminated, but even if some sources of feed were sterile, rodents or other means could have

> **Box B**
> **Cysticercosis**
>
> Cysticercosis is widely acknowledged to be a disease of poverty, particularly where small-holder farming includes raising pigs that roam freely and where sanitation is poor or non-existent and thus pigs have access to human waste. It is caused by the cestode, *Taenia solium*. Humans are natural definitive hosts and acquire infections by consuming larval cysts in undercooked pork. Pigs are intermediate hosts and acquire infection by consuming eggs shed in human faeces.
>
> Latin America, India, Nepal, south-western China, Viet Nam, Indonesia (Papua Province) and eastern and southern Africa are the main foci for cysticercosis infections. In poor, rural areas, pigs are raised free-range and there is little quality control and inspection of carcasses prior to consumption. Pig farming and pork consumption are increasing in southern and eastern Africa, where unregulated slaughter still predominates. Surprisingly, compared with information on the prevalence of *Taenia* cysts in pigs, data for human cysticercosis in southern and eastern Africa are scarce, suggesting a lack of understanding of the mechanism of *T. solium* transmission.
>
> Vaccines against cysticercosis are becoming available and it is targeted as an eradicable disease. Control programmes must concentrate on effective methods to disrupt transmission of *T. solium*. The Bill & Melinda Gates Foundation has recently funded intervention trials for cysticercosis in South America. Intensification of pig farming in sub-Saharan Africa may reduce the incidence of cysticercosis, even in the absence of improved hygiene regulations and their enforcement (69).

spread the salmonellae (68). Vertical salmonella transmission (from chicken to egg) is also important.

Foodborne disease can also occur if animal faeces that contain pathogens such as *Escherichia coli*, *Campylobacter* spp. or *Salmonella* spp. contaminate muscle or organs after slaughter in the abattoir. Some zoonotic diseases, such as Q fever and anthrax, are especially associated with abattoir workers. Food may also become bacterially contaminated at the processing stage; for example, from contact with equipment or surfaces or during preparation from a microbially contaminated surface, knife or hand.

Adequate cooking can kill some pathogens, such as bacteria and helminths, but not prions nor the toxins produced by some bacteria and fungi. While the transmission of prions through ritualistic cannibalism has been recognized for several decades, a more recent example is variant Creutzfeldt–Jakob disease (vCJD), a foodborne disease associated with intensive farming methods (6).

Food can also be contaminated by mercury, arsenic (from naturally contaminated irrigation water), pesticides and other agrochemicals. Mycotoxins, including aflatoxins, are naturally occurring toxicants produced by fungi, including *Fusarium* spp. Periodically, multiple deaths have been attributed to acute aflatoxin poisoning in East Africa and India (*70*) and mycotoxin-contaminated barley is suspected of being an etiological agent for Kashin–Beck disease (*71*).

Aflatoxin contamination of foodstuffs can be a major problem for impoverished people experiencing food insecurity. Because people eating such contaminated food will usually not see or taste anything abnormal, quantities sufficient to cause later problems, including liver failure and hepatocellular carcinoma, can easily be ingested (*72, 73*). Although a simple fluorescence test can usually determine aflatoxin contamination, it is not widely used in developing countries. Climate change may well increase the burden of disease from mycotoxins, including through increased rainfall events during or immediately after harvest (*74, 75*).

Ciguatera and certain shellfish secrete naturally occurring toxins (*76*), with ciguatera thought to cause more than 50 000 cases of poisoning per annum (*77*). Naturally occurring plant toxins in legumes, cassava and yams are also responsible for diseases such as konzo and lathyrism, which affect tens of thousands of people in poor countries, a problem far larger than previously estimated (*78, 79*).

Microbiological contamination of food plants, especially those that are harvested close to the ground and which are eaten uncooked, such as strawberries and many forms of salad greens, occurs particularly via animal or human faeces (*80, 81*).

3.7 Diseases transmitted by body fluids

Many diseases are transmissible by the exchange of body fluids and tissue, such as blood, breast milk and saliva and via organ transplants, medical procedures and sexual intercourse. Important within this category are many viral diseases, such as HIV/AIDS and viral hepatitis B and C. Increasingly, blood banks outside of Latin America are screening for Chagas disease due to its rising rate within immigrants to Europe and North America. Rabies is occasionally transmitted via organ transplants, including corneal. This risk is enhanced because the incubation period of rabies can be prolonged, and the organ donor may be asymptomatic at the time of death (*82*). Variant CJD may also be transmissible by blood transfusion (*83*). Particular zoonoses, such as brucellosis, are transmitted through direct contact of humans with animal fluids.

3.8 Other possible classifications of human infectious diseases

3.8.1 Socio-economic status

While many infections are clearly associated with poverty, some (e.g. legionellosis, bovine spongiform encephalopathy (BSE) and infections associated with type-2 diabetes) are commoner in more affluent populations.

3.8.2 Vaccination status or underlying immune status

Cryptosporidiosis and Karposi sarcoma (caused by human herpesvirus 8) are commoner in severely immunocompromised populations, such as those with HIV/AIDS or among recipients of transplants. Viral diseases transmitted via transplants may also become activated more quickly in the recipient than in the donor, as discussed above for rabies (*82*).

3.8.3 Vaccine preventability

Cholera, influenza, Japanese encephalitis, measles, poliomyelitis and yellow fever are vaccine preventable. Progress is being made towards developing a vaccine for dengue fever (*84*). Work has been carried out for decades on malaria vaccines, with hope for success having recently being revived (*85*). Vaccines are being developed to lower the burden of disease in animals that can transmit zoonoses, such as *Taenia solium* in pigs (*86*). However, despite years of effort, many important diseases lack an effective vaccine, including malaria, leprosy, schistosomiasis and HIV/AIDS. The course of some diseases, most importantly TB, can be modified by vaccination, though not fully prevented (*87*).

3.8.4 Form of the infectious agent

Pathogens can be classified as viral, bacterial, protozoan, fungal, helminth (cestodes, nematodes and trematodes) (*88*), insect or prion. Several transmissible forms of cancer exist; for example in Tasmanian devils (*Sarcophilus harrisii*), dogs and some immunosuppressed humans (*89*).

3.8.5 Zoonoses, reverse zoonoses, anthroponoses and epizoonoses

In 1959, a WHO Joint Expert Committee defined zoonoses as those diseases and infections that are naturally transmitted between vertebrate animals and humans (*90*), but they can be classified further (*91*). For example, some pathogens that were originally zoonotic, but which can now circulate within humans without requiring an animal host (e.g. influenza virus, measles virus, variola virus and more recently human immunodeficiency virus (HIV)) are sometimes called 'old zoonoses'. In contrast, pathogens that either originated in humans or evolved into forms that do not require any other animal as host, but which can opportunistically infect other animal species, are termed 'reverse zoonoses' (previously termed 'anthropozoonoses') (*92*).

TB was long thought to have originated in an animal; in fact, it may be a reverse zoonosis (*93*), as may be the strain of H1N1 influenza virus that recently caused swine 'flu (*94*). On the other hand *Mycobacterium bovis* may have evolved in a third non-human mammal and spread to cattle, humans and many other species (*95*).

Pathogens that are maintained exclusively within humans, without other animal reservoirs, are sometimes called anthroponoses (e.g. *Plasmodium falciparum*, causing malaria, or *Onchocerca volvulus*, the parasite that causes onchocerciasis). The south Asian form of visceral leishmaniasis is generally considered to be an anthroponosis; however, since it can be transmitted to other species, including cattle and goats, it may also be classed as a reverse zoonosis. Nevertheless, it has recently been shown (using the polymerase chain reaction (PCR)) that persistent leishmania DNA can occur in goats in Nepal, suggesting that these animals may be involved in the transmission cycle (*96*). In the New World, reduced biodiversity, deforestation and urbanization have been hypothesized to be triggers of evolutionary forces and may be changing the epidemiology of leishmaniasis from an exoanthropic to a synanthropic zoonosis (*97*) (pathogens that circulate only between animals and humans that live in close association) or even an anthroponosis (*98*).

Dengue fever is maintained within humans, without need for any other animal host (though transovarial transmission occurs in some mosquito populations), although parallel sylvatic cycles in primates have been documented in transmission foci in West Africa and peninsular Malaysia (*99*). This cycle occasionally spills over into humans. Epizoonoses refer to pathogens, such as those that cause rinderpest and bluetongue disease, that circulate exclusively within non-human vertebrate species, and which are discussed below.

3.8.6 Burden of disease over time

The burden of disease due to plague has been immense over the past two millennia (*41, 50*) but was minor in the twentieth century. Diseases involved in the 'Columbian' and other forms of exchange (*100*), such as measles, smallpox and possibly leptospirosis (*48*), have also exerted an immense burden of disease over time, decimating populations and altering the course of civilization.

Today, the neglected tropical diseases — including Chagas disease, leprosy, onchocerciasis, schistosomiasis, and soil-transmitted helminthiasis — have a combined burden of disease at least equivalent to that of malaria and TB (*101*). The burden of disease of several conditions formerly considered as neglected, including poliomyelitis, filariasis and guinea worm disease, has been lowered.

3.9 Infectious diseases of non-human species that indirectly affect human health

3.9.1 Farmed mammals, birds and fish

Many infectious diseases of non-human species do not cause direct human harm. Most zoonoses are spread to humans from mammals and birds, though reptilian pets have occasionally spread *Salmonella* spp. to humans (*102*). Fish-borne zoonotic trematodes are important emerging and re-emerging pathogens that cause liver and intestinal fluke diseases in humans (*103, 104*), while the evolution of prion diseases in farmed fish and their consequent transmission to humans is only a remote possibility (*105*). Diseases that infect plants are not known to infect humans; however, many have a strong influence on human health and well-being by causing loss of livestock, livelihood, income and morale as well as having a negative effect on nutrition (*106*).

Many social and environmental changes influence infectious diseases in other species (*107*). Good examples include bluetongue disease, a climate-sensitive, vector-borne viral condition (affecting sheep and cattle) (*108*), and rinderpest (see Box C) (*109*).

Box C
Rinderpest, the Maasai and social-ecological complexity

Mota and Qaranyo, why are they not ploughed?
I came from there to here without seeing an ox.
– Line from Ethiopian poem (*109*).

In the late nineteenth century, the inadvertent introduction of rinderpest (an epizoonosis) in imported cattle led to catastrophic ecological and human health effects throughout Africa (*109*). In Ethiopia, more than 90% of cattle died and loss of domesticated oxen reduced ploughing and thus agricultural productivity. Infection of wild species lowered the effectiveness of hunting; while lions, deprived of their usual prey, turned to eating humans. Exacerbated by periodic droughts, as many as one third of the Ethiopian population and two thirds of the Maasai people of East Africa died. This period is still remembered by the Maasai as the *Emutai* (meaning 'to wipe out') (*110*).

Social fragmentation also resulted, with increased raiding of stock and crops (*111*). Massive loss of herbivores altered the grassland ecology, allowing more tsetse-fly-bearing thickets to grow, thus increasing mortality from African trypanosomiasis (*110*). Throughout history, rinderpest has been considered the most important disease of cattle, and is probably the origin of measles (*112*). Controlling rinderpest was an early priority of the Food and Agricultural Organization of the United Nations (FAO) and it is believed to have been eradicated worldwide in 2010. It thus joins smallpox as the second disease that has been eradicated in the wild.

Foot-and-mouth disease is an animal epidemic that periodically has a negative effect on human well-being, although it only rarely causes human illness and does not directly kill many animals. Rather, most harm to humans is through trade restrictions since, in order to safeguard exports, it is often considered necessary to slaughter affected cattle, many of which are either well or likely to recover (*106*).

Sometimes responses to zoonoses intended to minimize human infection inflict substantial damage to human well-being. Recent examples are the slaughter of infected and potentially infected pigs in Malaysia and the Philippines due to outbreaks of Nipah virus infection and another, initially suspected as the species *Reston ebolavirus*, respectively. In 2009 a mass cull of pigs in Egypt was instigated due to fears of H1N1 swine 'flu, even though no Egyptian pigs had been infected (*113*).

Epizootics in farmed fish can also compromise human well-being. For example, in 1995 and 1998 a newly discovered herpes virus in wild pilchards spread for thousands of kilometres along the Australian coast, causing multi-million dollar losses to the fishing industry. The origin of this outbreak was suspected to be infected imported frozen pilchards that were used to feed penned bluefin tuna (*114*). More recently, it has been predicted that an outbreak of infectious salmon anaemia reduced the 2010 Chilean farmed salmon harvest by more than 80% (*115*). Farmed fish are also vulnerable to eutrophication, such as from excessive fish feed. Coastal eutrophication, sometimes leading to 'dead zones' due to excessive fertilizer run-off (*116*) can also lower farmed fish production, by contributing to several diseases, both infectious and non-infectious (*117*, *118*).

The exponential growth of aquaculture in Asia has been suggested to be the most important risk factor for the emergence of liver fluke infections, e.g. clonorchiasis and opisthorchiasis, which are classified as major etiological agents of bile duct cancer (*103*, *104*). Until recently the dominant share of the aquaculture production in the liver fluke endemic areas has been eaten within the communities near the freshwater bodies where the fish have been cultivated; however, this is changing with improved transportation and distribution systems. Thus it is anticipated that the spatial distribution of fish-borne liver fluke infections will expand to areas where no fish are being farmed (*104*).

3.9.2 Birds, bats, bees and amphibians

Infectious diseases of wild birds, bats and bees can also harm human health. Wild birds can be infected with strains of avian influenza, and it is now accepted that wild birds play a role, albeit minor, in the huge geographical jumps that occur in the distribution of avian influenza, which primarily infects domestic chickens and ducks (*118*). However, a 2006 meeting jointly sponsored by FAO and OIE concluded that the commercial poultry trade is still the principal route for avian flu dissemination (*119*).

Wild birds also serve as host species for zoonotic viruses, such as vector-borne West Nile virus. In turn, such pathogens (especially if introduced to immunologically naïve bird populations (*120*)), can alter the distribution of avian species, with additional ecological effects — another aspect of pathogen pollution (*121*). A suggestive historical example is the extinction of the passenger pigeon (*Ectopistes migratorius*) in North America, largely as a result of hunting pressure.

Loss of this species appears to have contributed to the spread of Lyme disease in North America, because the resultant higher abundance of acorns (mast) led to a rise in the population of tick-bearing mice (*122*), which (in association with other ecological and behavioural changes) increased the incidence of this disease (*123*). This example illustrates loss of an ecosystem "disease regulating" service (*124*). Migratory birds can also carry the ticks that help disperse Lyme disease (*125*), although it is unclear whether passenger pigeons did so. The final item in the chain was that an invasive plant, *Berberis thunbergii*, increased the habitat for the ticks that carry *Borrelia burgdorferi*, the causative agent of Lyme disease (*24*).

Bats increasingly share the human ecological space (*126*) and deserve special mention because they are an important reservoir of emerging zoonotic viruses worldwide. It has long been known that they can transmit infectious diseases such as bat rabies directly to humans, but they can also transmit Nipah virus to humans by contaminating palm sap with infected fluids (*127*). Zoonotic transmission from bats may be indirect, involving domesticated and farmed species such as pigs (Nipah virus and the species *Reston ebolavirus*), horses (Hendra virus) and possibly civets and other small farmed mammals (SARS). For African Ebola and Marburg viruses, the link is considered strong but the transmission mechanisms remain unknown (*128*).

Humans and their domesticated species, such as dogs, appear to have been involved in the transmission of numerous epidemics among animals, including seals, manatees and turtles (*129*). Among land animals, humans appear to have contributed to the spread of white nose syndrome, an emerging fungal disease found in several species of bats — rapid spread of which in the north-eastern USA has been linked to recreational use of caves (*130*). Bats play critical, underappreciated ecological roles that are vital for agriculture, food security and thus human immunological resistance to infectious diseases, including insect control, plant pollination and seed dissemination.

Diseases involving bees, such as colony collapse disorder and varroatosis, and other pollinators also fit in this category (*131, 132*). Furthermore, the loss of amphibians and other visible and 'charismatic' species due to a complex of emerging infectious diseases (*133*), climate change and habitat destruction can reduce income from tourism and cause loss of the cultural services that such species supply (*124*). Existence or altruistic values may be especially relevant to people who have strongly expressed biophilia (*134*).

3.9.3 Infectious diseases of plants

The oomycete *Phytophthora infestans*, which caused the Irish famine in the 1840s, is still the world's most important threat to potatoes (*107*), while the fungus *Puccinia graminis* has recently re-appeared as a risk to the global wheat crop (*135*). Numerous other viral and fungal diseases threaten crops, as do many insects and other pests, such as the brown rice hopper (*136*). Their distribution is likely to be influenced by climate and other environmental changes.

Carbon 'fertilization' — the effect on the ecology of plants caused by increased levels of carbon dioxide (CO_2) — can sometimes lower plant defences. Examples include the Japanese beetle (*Popillia japonica*), a threat to soy bean crops, and a variant of the Western corn rootworm (*Diabrotica virgifera virgifera*) (*137*). Recently, cassava brown streak disease, a viral disease transmitted by whitefly vectors and infected plant material, has been identified at unusually high altitudes, consistent with the warmer temperatures associated with climate change (*138*). Infectious diseases of plants could thus have an important effect on food security, nutrition, immunity levels of poverty and indirectly, on the epidemiology of many infectious diseases.

3.10 Emerging infectious diseases

Emerging infectious diseases include apparently new infectious diseases (in humans, animals and plants) and known diseases undergoing a substantial increase in severity, virulence, spatial range or a combination of these. The development of antimicrobial resistance has long been considered to be a component of disease emergence (*7*). Insecticide resistance, which can also expand and revive the range of infestations (see Box A), such as the current upsurge of bedbugs and vector-borne diseases, has more recently generally been excluded as a form of emergence (*5*). Nevertheless, it warrants re-inclusion as a category of resistance, and would be consistent with some earlier formulations of the term (*139*). Helminth resistance in livestock may prefigure that in humans (*140*).

A major category of emergence is diseases that were previously entirely unknown. Some of the 335 emerging disease 'events' described by perhaps the most widely cited recent paper on this topic (*5*) are caused by pathogens that have been identified since 1940, but which are unlikely to be genuinely new; for example, Murray Valley encephalitis, viral hepatitis C and Ebola virus infection. Others, such as diseases caused by the Henipaviruses (Hendra virus and Nipah virus) are also old (at least in bats), but their occurrence in humans may be genuinely new — though there is probably a much longer history of Nipah virus in south Asia, even if unrecognized, than in Malaysia (see section 4.1). Even HIV is probably not entirely new to humans, though its global spread certainly is. Some emergent diseases, however, do appear to be completely new, at least in humans. The best examples are SARS and vCJD (see Table 3).

Research into the form of emergence arising due to drug and/or insecticide resistance has a lower priority than that into other forms of emergence. Although its exact circumstances cannot be predicted, such resistance appears to be an inevitable, evolutionary-driven phenomenon that was recognized and predicted by the pioneers of antimicrobial therapy (*141*).

Table 3
A classification of emerging diseases by disease and economic burden[a]

Form of emergence and re-emergence	Example(s)[b]	Human disease impact and burden (scale, 0–5)	Other costs (scale, 0–5)
Genuinely new human pathogen	HIV/AIDS	5	5
	vCJD	1	3
	SARS	1	3
	Nipah virus infection (Malaysia)	1.5	2
	Hendra virus infection	1	2
New strain of known pathogen	Spanish influenza	5[c]	5
Known pathogen in new population	TB, measles, smallpox (indigenous peoples)	5[d]	5
Newness uncertain	Nipah virus infection (South Asia)	1	1
Newness unlikely	African Ebola virus infection	2	2
	Guanarito virus infection (Venezuela)	2	2
Drug resistance	Malaria	3	3
	TB	3	3
Insecticide resistance	Malaria	3	4
	Dengue fever	2	2
	Kala-azar	2	1
Multi-causal	Dengue fever	3	3

[a] Emerging diseases are usually grouped together, but they differ substantially in their newness, causal category, disease burden and other costs. Re-emerging diseases due to insecticide resistance are omitted in most classifications; we argue that this is a conceptual omission. The burdens and costs suggested here are relative and reflect the *additional* burden due to the emergence. For example, the total cost of malaria is higher than that solely from drug and insecticide resistance. The diseases shown are not intended to be comprehensive.

[b] vCJD = variant Creutzfeldt–Jakob disease; SARS = severe acute respiratory syndrome; TB = tuberculosis.

[c] High impact, but burden short-lived.

[d] Historically.

Several factors are integral to the emergence of genuinely new infectious diseases in humans and animals. One is the change in frequency and intensity of contact between humans and a range of wildlife species, including some that are intensively farmed, such as civet cats (*142*). Forest clearance, recreational hunting and travel have distributed pathogens, as has trade in vertebrate species. Pathogen dissemination has also been stimulated by globalized trading patterns and by global climate change (*143*). Hydraulic engineering may also play a role, though this is more likely to trigger changes in the epidemiology of existing diseases, such as schistosomiasis (*144, 145*) and cryptosporidiosis (*146, 147*).

Although most genuinely new human infections do not have a very high disease burden, there are important exceptions, as exemplified by HIV/AIDS and the Spanish strain of influenza (*148*). Familiar human diseases in non-immune populations can also be considered to be emergent in their historical context, such as plague in Europe and smallpox and measles in the New World. Many emerging diseases also affect non-human animals and plants; some examples are discussed elsewhere in this report.

3.11 Infections and chronic diseases

If untreated, many infectious diseases have chronic manifestations (*149*). Chagas disease, for example, is a leading cause of chronic cardiac, neurological and bowel conditions in Latin America. Rheumatic fever often leads to valvular heart abnormalities that can significantly shorten life, and rheumatic heart disease is estimated to cause more than 200 000 deaths a year (*150*). Cysticercosis is a leading cause of chronic epilepsy in some developing countries (*67*). Infection with *Schistosoma mansoni* and *S. haematobium* are associated, respectively, with portal hypertension and chronic and sometimes malignant bladder disease. *Clonorchis sinensis* is also associated with cholangiocarcinoma, while malignancy and epilepsy are linked with a range of helminthic infections (*149*).

Stunting, usually caused by the interaction of poor nutrition and persistent infections in childhood, also causes chronic disadvantage (see chapter 7).

'Epigenetic' mechanisms (i.e. where there is no change in the DNA but yet heritable changes in the gene function occur) can be transmitted over several generations (*151*). Such mechanisms can be triggered through nutrition and food availability, including caloric and nutrient deprivation from conception to toddler stage.

4. Environmental and agricultural drivers of infectious diseases of poverty

Although the contribution of the environment to the global burden of disease has been estimated to be approximately 24% (*152*), such estimates depend on the 'conceptual lens' of the observer (*153*). For example, the environmental contribution to HIV/AIDS in the study cited in ref. *152* was considered to be due to occupational transmission by sex and migrant workers, even though HIV/AIDS is an old zoonosis that has repeatedly crossed into humans from bushmeat hunting (*154*).

The key environmental driving forces discussed in this chapter are deforestation, habitat fragmentation, ecological disruption and contamination, agricultural intensification, and climate change (see Figure 8) (*155*). Global environmental change interacts in numerous ways with human infectious diseases risk. Poverty amplifies almost all forms of infectious disease risk, including those diseases with an environmental determinant.

Figure 8
Interaction between global environmental change and human infectious diseases (IDs)

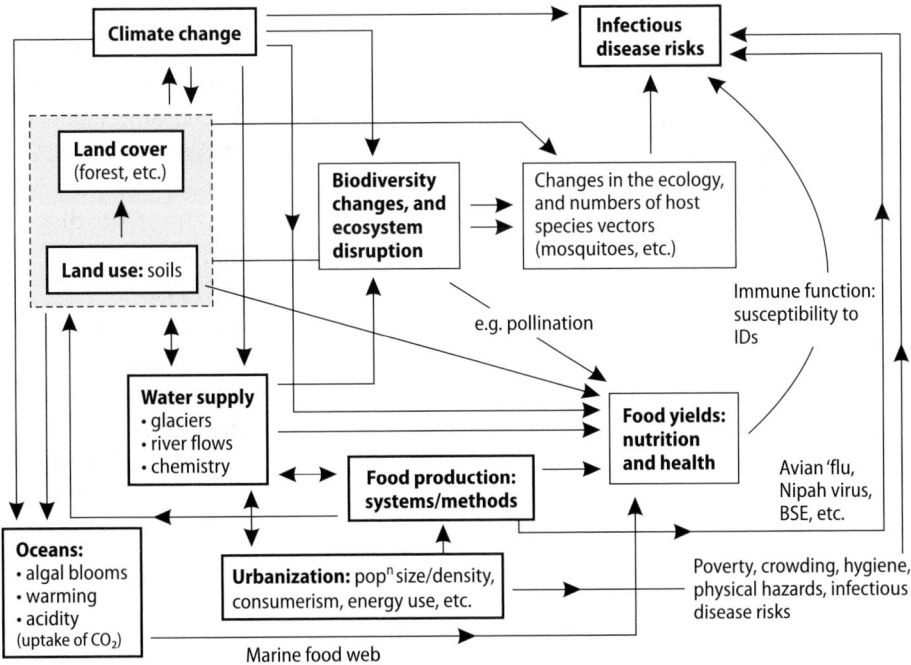

BSE = bovine spongiform encephalitis. Adapted from Figure 2.8.3, published in: McMichael AJ, Bambrick HJ. The global environment. In: Detels R et al., eds. *Oxford textbook of public health*, 5th ed. Oxford, Oxford University Press, 2009, by permission of Oxford University Press (www.oup.com).

The social drivers, including poverty, population growth, urbanization and cultural forces and institutions, all of which form part of the pathogenic 'milieu' framed by Claude Bernard (*156*), are discussed in chapter 5.

4.1 Forestry changes, ecological disruption and contamination

Habitat fragmentation, species loss and global toxification are largely attributable to human activities. Deforestation, which reduces habitat and biodiversity, and is a major contributor to climate change (*157*), occurs in many developing countries mainly to create land for agriculture and plantations, including biofuels. Though its net effect has been a larger, more prosperous human population, deforestation cannot continue indefinitely. Although a recent FAO assessment found that annual global forest clearance declined by about 20% in the 2000s over that in the 1990s (*158*), this trend is unlikely to continue, especially because of pressures to enhance food security (*159*).

Tropical deforestation contributes to the risk of occurrence of emerging infectious diseases, but not all geographical areas present identical risks. 'Hot spots' have been hypothesized, where the collocation of animal species, human behaviour and pathogens capable of jumping species amplifies infectious disease risk (*154, 160, 161*).

Deforestation also alters soil run-off and changes the chemistry and sediments of water courses, which may affect species-specific requirements and alter pathogen–host density. It can also increase flooding, especially when the ground is already waterlogged towards the end of the wet season, although this relationship has been challenged (*162*); nevertheless, the devastation caused by Hurricane Mitch in 1998 (at the end of the rainy season) was intensified in deforested areas (*163*). Flooding can also influence infectious diseases directly by causing undernutrition (*164*), physical trauma and infection; for example, through exposure to infected rodent urine in flooded fields, leading to leptospirosis (*165–167*).

Deforestation may also increase the habitat for disease-transmitting species; for example, hantavirus-bearing rodents, whose numbers may multiply not only due to loss of larger predators (*47*) but also of competitors. It can also provoke increased contact between humans and displaced species, e.g. bats, elephants and monkeys. Forest roads can partition habitats and create edge effects and enhanced fire risk, due to drying and risk of accidental or deliberate ignition. They also facilitate bushmeat hunting, with its attendant health risks and benefits. But roads also facilitate trade; and expose people to new influences, new ideas and contact with others. Their health effects are likely to be complex (see Box D).

> **Box D**
> **Diarrhoea, new roads, and social networks in Ecuador**
>
> An interdisciplinary research group started fieldwork in 2003 on a project in Esmeraldas Province in Ecuador to investigate the long-term health effects of a newly paved road into the area.
>
> Of 120 villages in this province, 21 were randomly chosen for study. Some were located on the paved road, while others were accessible only by dirt logging roads or river. Oil palm plantations and deforested hillsides are expanding each year in this region. Roadside squatter settlements have become wooden towns, and cellphone towers, buses to the beach, and fibreglass boats have appeared and are used concurrently with palm leaf huts, horses and wooden dugout canoes. The study examined the variability in these changes across space and over time, as well as their links to health status, especially for diarrhoeal diseases, nutritional status and dengue fever.
>
> The study team has visited each village every six months since 2003, gathering demographic, geographical and epidemiological data. In addition, full-time ethnographers circulate regularly, interviewing and observing in each village. This approach permits analysis of disease transmission at household, village and regional levels, which is critical to understanding and differentiating individual and system-wide influences.
>
> Results show that remoteness from a metropolitan centre is negatively associated with episodes of diarrhoeal disease (*168*); social network density and household proximity jointly predict risk of disease within villages (*169*); and social network densities across rural villages markedly influence the likelihood of disease transmission (*170*).
>
> This region, the Gran Chaco forest, is one of the world's most important reservoirs of biodiversity (*171*). Analysts have described deforestation and other ecological insults in the region (*168*, *172*), but without reference to human health. Interdisciplinary approaches to studying disease transmission are critical because they help link regional and temporal changes in demographics and epidemiology to both human motivations and microbiological changes.

Infections associated with forest clearing include vector-borne haemorrhagic fevers, such as Venezuelan haemorrhagic fever, Mayaro fever and that caused by Junin virus (see Table 4). However, good local governance and economic development can reduce the risk of contracting such conditions (*173*). In the Amazon, variations in the risk of contracting malaria have been linked to forest clearance and agricultural settlement. A combination of factors thwart malaria control programmes in newly deforested regions, with the notable exception of corporate-sponsored forest clearance — here, the landscape was transformed faster, the personnel involved knew more about the risk of malaria and took better protective measures against mosquito exposure (*174*).

Table 4
Geographically restricted viruses[a]

Country/area	Common name
Cameroon, Nigeria	Lassa fever virus
Southern Africa	Lujo virus
Venezuela	Venezuelan haemorrhagic fever (Guanarito virus)
Bolivia	Machupo virus
Argentina	Junin virus

[a] Some newly discovered diseases are geographically restricted, perhaps because they lack the characteristics and ecological context to be widely disseminated. Some others (e.g. HIV/AIDS, Chikungunya virus disease, dengue fever) do have the capacity for such spread.

Many factors associated with the recognition and emergence of infectious diseases remain incompletely understood. While the tropics are the greatest source of such diseases and have the greatest mammalian biodiversity, these two elements are poorly correlated. For example, although Indonesia has the highest mammal diversity in the world, few emerging infectious diseases have been identified there.

There is an interaction between deforestation, pressures on wildlife and the intensification of animal farming, with the most obvious mechanism being that these factors increase contact between wildlife and humans (*34, 126*). Altered pathogen dynamics within host populations that enhance virulence may also contribute (*17, 34, 175*), as may increased viral shedding from wildlife due to stress (*176, 177*) (see Box E).

Several outbreaks of Nipah virus infection have been reported in Bangladesh and nearby north-east India. Initial reports primarily concerned young boys who had been exposed directly to infectious bat droppings, perhaps by playing under trees where the animals were roosting. More recently, spread of the virus has been linked to the ingestion of date palm sap that had been contaminated by the urine and faeces of fruit bats (*127*). Both explanations are plausible, with the latter being another form of linkage between food harvesting and infectious disease risk. Person-to-person transmission of Nipah virus also occurs, and some cases with no known risk factor have been recorded in south Asia (*181*). Other ecological factors have been attributed to the emergence of Nipah virus and perhaps other bat-transmitted diseases (see Box E). Deforestation has also been linked with schistosomiasis and malaria (*1*) (see section 6.3).

> **Box E**
> **Does bat stress increase bat vulnerability to infection and viral spillover?**
>
> In parts of south-east Asia much fruit bat habitat has been replaced by plantations such as oil palm, grown for food and biofuel (*178*). Such plantations are poor habitat for bat species other than *Cynopterus brachyotis*, the lesser short-nosed fruit bat, thus increasing the chance of most bat species feeding in orchards and urban settings (*179*). Deforestation, mainly from burning, leads to significant seasonal atmospheric haze, which reduces flowering and fruiting of forest trees, further changing the natural food of fruit bats and possibly altering their migration patterns and ability to find food. These synergies may be causing stress to bats, possibly altering their viral loads (*109, 179*).
>
> The occurrence of the bat-borne Nipah virus outbreak in Malaysia during the strongest recent El Niño event (1997–98), which enhanced drought and thus fires and haze, led to speculation that the fires were associated with stressed, dislocated bats (*180*). However, the amplifying role played by pigs eating fruit contaminated by contact with bats was quickly recognized (*180*). Subsequently, it is has found that this virus is widespread in south and south-east Asia, and public health measures have successfully prevented a recurrence in Malaysia. Cases of infection with Hendra virus, a species closely related to Nipah virus, have increased recently in north-east Australia in both horses and humans with close equine contact; it is thought that the essential mechanisms driving this are ecological (*126*). Spillover events generally occur in the dry season and have been reported since 1994. Adoption of urban habitats by bats and reduced migratory behaviour have also been suggested as causal factors in equine cases (*177*).
>
> Many populations of bats (a long-lived species) are probably stressed by anthropogenic habitat loss, unusual flooding, and by occasional deliberate attempts to relocate them. Some studies have found increased viral loads in bats that were associated with both pregnancy and nutritional stress (*176, 177*), perhaps related to immunosuppression (*179*). Reduced bat mobility may also lower herd immunity, with increased expression on viral re-exposure and re-introduction (*177*). The hypothesis that ecological changes alter viral spillover via immune effects should be further investigated.

4.2 Dams, lakes and irrigation systems

The harnessing of rivers for irrigation, flood control, and hydropower generation has brought many benefits. However, these benefits are also accompanied by harm to health, ecology and human rights, especially for people displaced by the flooding of densely populated tropical valleys in some of the world's poorest countries. Sometimes the harm done to such people can commence decades before dam construction, since it becomes irrational to construct infrastructure that will one day be flooded (*182*). Dams also reduce the flow of nutrient-rich silt, which can lower soil fertility and also contribute to subsidence and flooding of fertile river deltas (*183*).

Even well-designed water, dam and irrigation systems can enhance or enlarge the habitats of disease vectors (e.g. malaria-transmitting *Anopheles* spp. mosquitoes) (*184*) and of intermediate hosts such as the freshwater snails that transmit schistosomiasis (*185*). Dams also disrupt natural hydrological ecosystems and water-filtration processes.

Disruption to lacustrine predators and the resultant ecological changes in lakes have also been linked to changes in the epidemiology of schistosomiasis. Evidence from Lake Malawi, for example, suggests that overfishing of mollusc-eating fish (cichlids) has increased the numbers of *Bulinus* gastropods and the subsequent spread of *S. haematobium* (*186*). In Cameroon and Kenya, the introduction of cichlids has been relatively ineffective in eliminating or reducing schistosomiasis, but there may be a role for other natural predators in integrated control efforts. Research into the relationship between snail species diversity and schistosome transmission may reveal other opportunities for controlling transmission through protection or manipulation of snail communities (*187*).

A meta-analysis of 58 studies found that about 779 million people are at risk of schistosomiasis, 106 million of whom (13.6%) live in irrigation schemes or in close proximity (< 5 km) to large dam reservoirs. Importantly, occurrence of *Schistosoma mansoni* was 2.5-times higher among people living in close proximity to large dams compared with those living further away; for irrigation systems the risk was almost 5-fold. It was concluded that strategies to mitigate schistosomiasis should become integral parts in the planning, implementation and operation of future water projects (*188*).

4.3 Agricultural intensification

Growing populations, increasing average incomes (*189*) and the emerging global consumer class (*190*) have generated an enormous increase in intensive farming of both plants and animals (*191*). Like deforestation and dams, the net effect of agricultural intensification on human health is beneficial (*192*); the issue is to what extent it can be made more sustainable (*192*) and expanded (*159, 193*).

More reliable and larger food supplies also enable denser human settlement, conducive to the maintenance of infections. This ancient process is still occurring, as the conversion of wild ecosystems brings humans into new forms of contact with other species, both directly and through intermediary species such as pigs or horses.

Intensification of animal farming may facilitate new zoonoses, such as highly pathogenic avian influenza and H1N1 influenza (*194*) (see Box F). It might also have been a factor in the emergence of SARS (via civet farming) (*195*) as well as of Nipah virus in Malaysia (*34, 175*), either the species *Reston ebolavirus* or another (*196*) (pig farms), and perhaps could be in future outbreaks of bat-borne diseases in Africa or South America (*197*).

> **Box F**
> **Agricultural intensification may be driving increased pathogen virulence**
>
> Intensive animal farming practices may select for fast-growing, early transmitted and more lethal parasites because the dynamics that determine the transmission–virulence trade-off in pathogens differ in artificial and 'natural' ecosystems.
>
> Most **intensive animal farms** are characterized by fixed location, high animal density, and homogenous, immunologically similar populations (*175*). Diets are likely to be repetitive and may lack micronutrients. Many animals have close contact with faeces; their own as well as those of other animals. Discarded feathers, dead animals (especially if birds) and packaging materials may act as plentiful fomites. Indoor farms have little or no exposure to sterilizing sunlight. In contrast, animals in **wild ecosystems** have high movement, lower densities and more immunological and ecological variation.
>
> Viruses that have evolved to survive in intensively farmed conditions are unlikely to be as well-adapted to wild ecosystems. For example, highly pathogenic avian influenza (HPAI) generally occurs and is sustained only in intensively farmed conditions, while among wild bird populations the virus usually exists in its more benign low pathogenic form (*198*).
>
> There is concern that HPAI may cause a human pandemic, but some have argued that if human-to-human transmission were to occur it might involve a trade-off with lower human lethality (*199*). Others have dismissed this view (*200*) and the recent experiments involving serial passage of virulent H5N1 to ferrets may also refute it, though they may not be a good model for human infectivity (*201*).
>
> However, the chance of an extreme human H5N1 pandemic may be less likely because humans live in ways closer to that of a wild population than to an intensive farm. Nevertheless, the highly abnormal harsh, crowded and undernourished conditions on the Western Front in the First World War may have contributed to the emergence of the highly pathogenic Spanish influenza (*148*).
>
> As urbanization increases, especially if accompanied by intense poverty, the living conditions for billions of people could increasingly mimic those of intensively farmed animals, which could drive new infections with the potential for high rates of transmission, morbidity and even mortality. Such a scenario could be worsened by a decline in public health capacity, including provision of nursing care, isolation and treatment.

Introduced crops can also facilitate the emergence or spread of infectious diseases. Apart from irrigation as a frequent source (see section 4.2), other examples include the consequences of introducing palm oil plantations in Latin America (see section 8.1.3.1) and a malaria epidemic associated with new cacao plantations in Trinidad in the 1940s (*147*). In Thailand, cassava and sugarcane cultivation reduced the density of *Anopheles dirus* but facilitated the breeding of *Anopheles minimus*, with a resulting surge in malaria.

4.4 Climate change and infectious diseases of poverty

Though many environmental issues are serious (*202*), the issue of climate change is perceived as one of the most pre-eminent of our time (*203*). So far, this concern has not been because of its impact on the epidemiology of infectious diseases; rather it has been about its likely effects on food and electricity production, water security, ocean acidification and sea-level rise.

The frequency of extreme weather events has long been predicted to increase because of climate change, and evidence is mounting to suggest that this is now occurring (*75, 204–206*).

The multiple health risks of climate change are slowly being appreciated (*207–209*). These extend well beyond the risks from heat stress, intensified air pollution, more severe flooding and other weather disasters: sea-level rise and regional food insecurity are likely to dislocate populations and in turn heighten the risk of conflict (*210, 211*). It is also likely to have profound, largely adverse effects upon agricultural, marine and other aquatic productivity, including the distribution of agricultural zones (*212*).

Changes in climate and ecology are and will increasingly directly alter the transmission of many infectious diseases (e.g. see sections 3.4, 3.5, 6.2, 6.4, 6.5, and 8.2), with vector- and waterborne diseases being likely to produce the greatest burden of disease (see Figure 7). By modifying the survival and reproduction rate of vectors and pathogens, and by altering vector activity, changes in climate and ecology could extend the endemic area or length of the transmission season (*213, 214*), although transmission may become more difficult in some cases (*28, 215*). The epidemiology of many categories of infectious diseases is also likely to be altered by climate change, including those borne by soil, rodents, food and vectors. Fungal diseases may also become more widespread (*65, 216*).

4.5 Other environmental and agricultural driving forces

Many other environmental driving forces interact with infectious diseases of poverty; for example, indoor and outdoor air pollution and dust which, for example, increase the risk of respiratory infections (*217*). The projected rise in energy costs that will occur when demand for fossil fuel energy far outstrips supply is also likely to have a large effect upon such diseases, by increasing the cost of transport, food, fertilizers, and household energy (*218–220*). The extent to which recovery of gas trapped in shale and coal formations will expand fossil fuel reserves remains uncertain. Combustion of this gas also contributes to climate change, though probably less so than that of oil or coal — a claim that is, however, contested (*221*). Recovery of such gas may also contaminate (*222*) and reduce supplies of ground water.

Increasing scarcity of phosphate is also likely to raise the cost of fertilizer and food, probably during this century (*223, 224*). Poor countries that need to import phosphates will be particularly vulnerable, though probably not for several more decades.

Human exposure to excessive amounts of agricultural chemicals can also affect the immune system, and hence susceptibility to infectious disease, and also cause some chronic diseases, especially in genetically vulnerable groups. Examples include Parkinson disease and some forms of lymphoma and possibly other cancers. Mycotoxins, which can occur in poorly stored grain, act synergistically with hepatitis B virus to increase the risk of hepatocarcinogenesis (72).

The use of antibiotics in livestock keeping has been conclusively linked to the development of microbial antibiotic resistance, resulting in more human disease.

5. Social drivers of infectious diseases of poverty

5.1 Poverty

Poverty is associated with many forms of vulnerability (*124*), both related to and arising from disease. Most of the burden of adverse environmental change in the coming decades is predicted to fall on poor populations in low-income countries.

There is a seemingly intractable cycle connecting vulnerability and poverty, exacerbated by the burden of infectious diseases, to which adverse environmental change increasingly contributes. For example, the poor are more vulnerable to the impact of extreme weather events because of factors such as flimsy housing, proximity to risks, and inadequate early warning systems. Poverty can also reduce agricultural productivity, through its strong negative linkages

---

Box G
The complexity of health promotion in Ethiopia

The economic and social history of Europe, Japan and North America show that health, capacity and other aspects of well-being are likely to benefit when the determinants of health (including the position of women) are improved. Of note also is that in the southern Indian state of Kerala, where there have long been high literacy rates and comparatively high status for women, life expectancy is far higher than would be expected from its low average per capita income (*228*).

The effectiveness of foreign aid might therefore be enhanced by strategies that seek to enhance human capacity, facilitating 'bottom-up' approaches in which populations in low-income countries become agents of change, generating 'virtuous circles' of improved health, education and governance. An example is an action-oriented approach in Ethiopia to enhance human capacity among thousands of pastoralists (*226*). Its aims were to diversify livelihoods, improve living standards, and enhance livestock marketing; included were efforts to improve literacy and numeracy to enable rudimentary banking and bookkeeping procedures, microfinance and other forms of collective action. The investigators observed a cascade of benefits and commented on how poor women became leaders and rapidly changed their communities.

However, this population may have been more replete in micronutrients than similarly poor populations in villages in India or in slums in other parts of Africa. The investigators do not comment on whether the population carried a high burden of parasites or of infectious diseases. Any such disease burdens, where they exist, are likely to limit the effectiveness of the interventions they describe.

Other work in Ethiopia has shown that simple interventions can harm health and development. For example, provision of wells closer to dwellings lowered the caloric expenditure of women, by reducing the time and effort they spent in fetching and carrying water. Child survival also improved; however, birth intervals were reduced and infants suffered from worse nutrition and greater stunting (*229, 230*). These examples suggest that development also requires broad and systemic approaches (see section 1.1.1).

to nutrition and education; access to extension services and information as well as to agricultural inputs and markets; and to land tenure. Being poor is often accompanied by innumeracy, increasing the risk of becoming victims of usury and of being trapped in chronic debt (*225*). It is often also associated with low agricultural output, which directly reduces income and increases vulnerability.

Deliberate attempts to improve human capital have sometimes proven remarkably effective at improving many aspects of human well-being, including health (see Box G) (*226*). That improvements in human capital can lead to self-organizing escapes from poverty and poor health is demonstrated by the enormous investment made by many countries in education and the prominence given to it as a key Millennium Development Goal (*227*).

5.2 Population growth

The widespread and continuing reticence to discuss the contentious issue of population growth (*231*) has fostered the idea that the degree of such future growth is either pre-determined or unimportant. However, this taboo (*232*) may be starting to lift (*233*).

Many initiatives to slow population growth have been associated with human rights violations (*234*). However, the factors most associated with high fertility in low-income settings (e.g. poverty, female illiteracy and subjugation, high infant mortality and lack of access to health care and family planning services and technology) constitute a huge injustice. Reducing these factors would improve health, slow population growth and facilitate economic take-off (*235*, *236*). It would also help protect poor populations against the gathering risk of climate change (*237*), lower its degree (*238*) and slow deforestation and ecological disruption (*231*).

5.3 Urbanization

More than half the world's population lives in cities, an increasing number in poorly or totally unplanned slums (*239*). Urban populations in developing countries are growing at a faster rate than those of countries as a whole, threatening the health advantage that urban populations have generally experienced (*240*, *241*). Parts of some densely populated cities, including Bangkok and Jakarta, are at serious risk not only from sea-level rise but from subsidence, due to over-extraction of groundwater (*242*).

When it is well-managed, urbanization need not lead to a decline in environmental quality (*243*); however, many cities in developing countries face deteriorating environmental and social amenities. Urban crowding, exacerbated by poverty, provides numerous chances for the exchange of pathogens, including drug-resistant forms. Cities with limited health education and material resources can also provide many opportunities for mosquitoes and other vectors to breed. Poor sanitation, undernutrition and changes in mobility, mixing and patterns of sexual relations add to the risks of infectious diseases, whether by resurgence or emergence (*244*).

Rural populations in low-income countries also face many environmental health hazards, including indoor air pollution (*217*) and a seemingly perennial lack of water and sanitation. Indoor air pollution accounts for the deaths of at least 1.5 million people annually, causing almost 3% of the total global burden of disease (*245*).

Some infectious diseases also affect higher-income populations in urban settings, such as dengue fever in many cities in Latin America (*246*) and some wealthy parts of Asia, including Singapore (*247*) and China, Province of Taiwan.

5.4 Cultural forces and institutional change

The persistence of the vast burden of infectious diseases of poverty is a human rather than a natural phenomenon, since there is sufficient knowledge to greatly ameliorate it. Today, the global economic system is marked by higher inequalities, irrespective of how income is measured (*248*). Although humans can tolerate high levels of inequality, evidence is accruing (*249*) that we have a clear preference for social situations where inequality is constrained (*250*) to levels that are far more equitable than those that exist today. While evolution arises because of competition for limited resources, groups of individuals, including humans, also cooperate in order to maximize sustainability of those resources, thereby reducing inequality within groups.

High global economic, social and health inequality nevertheless persist. Two reasons have been proposed for this. First, the evolutionary experience of the world as a single economic unit is short compared with the long history of our species. Many people, especially the 'bottom billion' (those living on less than US$ 1–2 per day) (*251*), are not yet fully incorporated into the social dynamics of this unit. Few of the bottom billion understand these larger global forces, and even fewer have the capacity to change them (*252*). Affirmative action, including reducing the burden of infectious diseases of poverty, is essential if their position is to be improved. Second, few people in the richest billion understand or want to understand the circumstances of the bottom billion, thereby impeding the evolutionary forces that would otherwise probably generate more equity.

There is growing appreciation that changes to global institutions (including to norms and cultural practices) are vital for sustainability (*253*), for the promotion of global health and for lowering the burden of infectious diseases of poverty. Organizations such as WHO, the United Nations and the Bill & Melinda Gates Foundation are substantially motivated by these forces. Given enough time, a global society may evolve in which the scale of inequality converges rather than remaining at its current level. The shift to open-source publishing, including for health problems of relevance to the poor, also reflects and encourages trends towards more fairness *(254)*. On the other hand, the increasing number of authors who have to pay journals to publish articles could lead to publication of lower quality material, because of the inherent conflict of interest of this process — which has been likened to a possible 'vanity' press in the worst case *(255)*.

6. Selected recent scientific advances, insights and successes

6.1 One Health–One Medicine

Over the past decade there has been renewed interest in the concept of 'One Health' and its antecedent 'One Medicine'. Stimulus for this comes from the impacts of globalization on microbial mobility and from more rapid changes to habitats and to inter-species contacts (*143*). Since ancient times it has been appreciated that the state of animal health and of the local environment influence human health (*256*). The importance of connecting research findings, clinical experience and learning from the human and veterinary domains was clearly recognized in the nineteenth and early twentieth centuries but then largely neglected (*143*). Calvin Schwabe coined the phrase 'One Medicine', which was later expanded to 'One Health' (*257*). An early pioneer of the ideas in this concept was Rudolf Virchow, while William Osler later adopted them (*258, 259*).

This interdisciplinary symbiosis was largely eclipsed during the twentieth century, as the 'biomedical' paradigm began to dominate ideas, clinical practice and epidemiological research on human disease. One effect of cellular and molecular biology, and of the development of systematic institutionalized medical education and health-care systems in wealthy developed countries, was to increase the perceived distance of humans from the rest of the natural world. However, renewed interest in the One Health concept re-emerged in the 1990s, particularly due to the influential Institute of Medicine (IOM) (*260*), one of whose reports emphasized that recent changes in the patterns of human and animal contact have made conditions more conducive to global outbreaks of zoonotic diseases (*261*). Another manifestation of this renewed interest was the formation in 2006 of the International Association of Ecology and Health and publication of its journal, EcoHealth (www.ecohealth.net) (*81, 262*).

One Health concepts are increasingly relevant in attempts to understand, predict, respond to, minimize and (hopefully) prevent the next influenza pandemic, whether on a 1957- or 1968-scale (case fatality rate, ca. 0.1% (*200*)) or the far worse scale of the pandemic at the end of the First World War (*148, 263*). There is growing understanding that domesticated species (especially pigs) serve as viral mixers for influenza viruses of human, porcine and possibly avian origin (*148, 160, 264*). There is also concern that very large populations of pigs and birds, among others, contribute more lethal pathogens via 'epidemic enhancement' (*34*) (also see Box F).

Ecohealth is a term which, at its simplest, attempts to link ecology and human health (*265, 266*). In its broadest sense, ecohealth covers many dimensions of health beyond infectious diseases, such as nutrition, mental health and social justice. This formulation overlaps and interacts with the broad version of One Health, sometimes called "One World One Health".

It is important to not exaggerate the capacity of One Health or ecohealth approaches to solve global health problems, since no single conceptual framework can do so. Instead, coalitions are needed between, for example, individuals who primarily view the world through the lenses of inequity, chronic diseases, antibiotic resistance, vaccine susceptibility, food insecurity, food safety and genetics.

The trademarked phrase "One World One Health"[5] (http://www.oneworldonehealth.org/) is broader than even One Health. It signifies the interaction not only of human and animal health but also with environmental change and social and environmental justice.

6.2 Eco-biological mechanisms of interaction

Here we discuss several other interactions between environment, agriculture and infectious diseases of poverty, including the evolution of new ecological niches for pathogens and the interactions of climatic and other environmental and social factors with infectious diseases of poverty. Such interactions may be classified by scale, intensification of production and by changes to the distribution of infectious agents and their vectors. Changes in the degree, distribution, severity and seasonality of global undernutrition will also affect these relationships.

6.2.1 The opening of new 'ecological niches' for microbes

The rapid degradation of ecosystems worldwide has increased the rate of background species extinction by up to 1000-fold (*267*). Basic ecosystem services that sustain human life are increasingly being compromised (*124*). The economic and human health and welfare (*268, 269*) consequences of environmental degradation are therefore rising.

6.2.2 Global trade in bushmeat and its interaction with infections

Bushmeat trade is a global phenomenon involving the harvest of free-ranging species (*270*), including endangered primates. The sheer numbers of species and individual animals killed for direct human consumption contributes to the biodiversity crisis. More than 500 million kg of bushmeat are consumed each year in the tropics alone, up to six times the sustainable rate (*267*). Deforestation and habitat fragmentation reduce animal dispersal capacity and facilitate human access to forest interiors, including hunters (*271*), loggers, graziers and settlers. Fragmented forests are also more vulnerable to fires and loss of diversity (*272*).

---

[5] This concept refers to an interdisciplinary, cross-sectoral approach to addressing human and animal health, underpinned by environmental stewardship. It is a trademark of the Wildlife Conservation Society.

6.2.3 Severe acute respiratory syndrome (SARS) and other bat-associated infections

SARS, an atypical pneumonia caused by a coronavirus, is a genuinely new human emerging disease (see Table 3), probably caused by increased degradation and exploitation of natural environments. It is thought to have originated in 2002 in Guangdong Province, a rapidly expanding area of southern China (*273, 274*). This area still hosts a thriving market in bushmeat, but no subsequent outbreak of SARS has occurred there. The disease may have been spread to humans by Himalayan palm civets (*Paguma larvata*) and raccoon dogs (*Nyctereutes procyonoides*) sold for human consumption in wildlife markets (*275, 276*). More recently, a SARS-like coronovirus has been isolated in Chinese horseshoe bats (*Rhinolophus sinicus*), suggesting a possible bat origin for the virus, which was then transmitted to humans following amplification, mixing and evolution in these secondary hosts (*277*).

A similar phenomenon, involving a partially immune population of intensively farmed pigs, has been hypothesized for Nipah virus, first described in Malaysia in 1998 after the deaths of 105 humans in close contact with domestic pigs (*34, 278*). This severe febrile encephalitis is caused by a paramyxovirus that was transmitted via pigs to humans from fruit bat hosts (see Box E). The emergence of another paramyxovirus, Hendra virus, occurred following transmission to humans in extremely close contact with horses that had been infected by fruit bats (*Pteropus* spp.) — mainly because of encroachment of human agricultural activities into recently deforested areas (*126*).

Ebola and Marburg viruses also cause forest-associated diseases that are transmitted by bats and by close contact with infected patients. Although their mortality rates are high, the burden of such diseases remains minor (*279*) and they appear to lack the characteristics for widespread human-to-human transmission.

6.3 Environmental quality and the burden of infectious diseases

Human health is influenced by environmental degradation, including that of water and air (*269, 280*). This is nuanced by considerations of time lags, 'chronotones' (*281*) (periods when health may temporarily decline due to environmental change but then improve), poverty, other social factors as well as the presence of and exposure to pathogens. For example, life expectancy fell in Britain during the early Industrial Revolution, which coincided with a decline in state-sponsored smallpox vaccination (*282*). However, later in the nineteenth century British life expectancy increased, largely because of imported food and the activities of social reformers (*283, 284*), rather than through advances in scientific knowledge.

Millions of American Indians perished in the sixteenth and seventeenth centuries as a result of new microecologies introduced by foreigners; for example, several virulent diseases transported from Europe and Africa, possibly including leptospirosis (*41, 48, 100*).

There are many other links between environmental degradation and increases in the incidence and severity of human infectious diseases. For example, marked increases in mosquito-bite rates (*285*) and an escalating incidence of malaria (*286*) have been linked to deforestation (*287*) — which can also change the microclimate, increasing the risk of mosquito survival and disease transmission (*187*). Flood risk and severity can also increase with deforestation of primary forest (*268*) as well as with climate change (see section 6.4).

Many disease outbreaks, including cholera, dysentery, viral hepatitis A, leptospirosis, malaria and schistosomiasis, are more likely to occur after floods (*288, 289*). Humans displaced by floods may also be forced to live in areas where inadequate sanitation and temporary high-density living conditions promote disease (*290*). Flooding can also have negative effects on nutrition, even after controlling for poverty (*164*).

6.4 Climate, seasonality, environmental change, geography and infectious diseases

Until recently, little attention had been given to the influence of climate on infectious disease epidemiology. Much more is therefore known about the role of non-climatic factors in determining the incidences, case distributions and likelihood of epidemic outbreaks. Yet climate sets the parameters on where and when many infectious diseases can be transmitted.

The advent of human-induced climate change alters this context and more attention is now being paid to identifying and modelling how climate change is likely to affect infectious disease occurrence. It also encompasses the altered risks of infectious diseases in food species and the consequent risks of shortages and undernutrition (*107*).

Climate change will affect human infectious diseases in diverse ways. Bacteria in food and in nutrient-loaded water multiply faster at higher temperatures. Changes in rainfall patterns affect river flows, flooding, sanitary conditions and the spread of diarrhoeal diseases. A shift to more irregular rainfall could alter vector population dynamics, leading to lower mosquito populations and transforming areas where malaria is holoendemic to a more epidemic pattern (*215*). Historical records in southern India show that cholera outbreaks are most likely either during times of drought or of flooding and crowding (*291*).

Many vector-borne infections are sensitive to temperature, rainfall, humidity and wind. With rising temperatures, mosquitoes feed more often, while pathogens within them (e.g. *Plasmodium* spp. and dengue virus) mature more quickly (*214*). Surface water patterns influence the breeding of mosquitoes and also that of the intermediate host snails of zoonoses such as schistosomiasis and fascioliasis (*292*); humidity affects mosquito survival (*293*). Soil characteristics, such as moisture, porosity and pH influence snail survival; while bamboo stands appear to be natural barriers to *Oncomelania* spp., the snails that transmit *Schistosoma japonicum* (*294*). Zoonoses are often influenced by climate-related changes in density and movement of the 'reservoir' animal species. Examples include birds and West Nile fever (now in Canada and the USA); cattle and Rift Valley fever (Kenya); and kangaroos and Ross River fever (Australia) (*295*).

Recently, several vector-borne infections appear to have increased their geographic range due to regional warming and other factors. These include malaria in highland areas (*296*), so far best documented in eastern Africa (*297*); tick-borne encephalitis and bluetongue virus disease in livestock in northern Europe (*102*, *298*); Crimean–Congo haemorrhagic fever in parts of southern and central Europe (*299*); and Lyme disease in Canada (*300*). With the recent introduction of *Aedes albopictus* into Europe, Chikungunya virus disease or even dengue fever may become established there (*301*).

Models that analyse the relationship between the timing of the onset of the monsoon and spring tides on the abundance of the adult mosquitoes that transmit Ross River virus in northern Australia over a period of more than 15 years were recently able to predict the magnitude of seasonal peaks in this abundance (*302*). This is important because increases in future intense rainfall (*303*), combined with higher expected frequencies and intensities of high tides, may increase the severity and perhaps frequency of disease-carrying mosquitoes in the tropics. For example, more rain will increase the availability of ephemeral freshwater breeding pools, favouring species such as *Culex annulirostris*. A greater frequency of high tides will increase suitable breeding habitats for *Ochlerotatus vigilax* and other saline breeding species, which may increase the frequency and transmission of mosquito-borne diseases in northern Australia (*302*). Similar models may help predict the timing and severity of meningococcal meningitis epidemics in Africa (*304*) and improve control and response programmes.

6.5 Climate change and helminthiases (other than schistosomiasis)

Among the helminths, climate change is most likely to change the epidemiology of diseases transmitted by trematodes, including fascioliasis (*88*, *305*), which has a broad geographic and altitudinal distribution and which is becoming increasingly problematic in parts of Latin America, Africa, Europe and south-east Asia.

The climate sensitivity of other helminths may have been underestimated (*305*). The distribution of *Echinococcus multilocularis*, the causative agent of human alveolar echinococcosis, appears to be more sensitive to climate change than that of *E. granulosus*, which can survive for months in a wide range of temperatures (*88*). Increased warming and lower humidity are thus likely to reduce survival of *E. multilocularis*, while increased rainfall could enhance it; in contrast, deforestation, other changes in land use, as well as other environmental factors would increase its distribution (*306*). Warmer temperatures in the Arctic and sub-Arctic regions could also expand the range and populations of foxes and voles, common carriers of *E. multilocularis* (*307*).

6.6 The value of the socio-ecological perspective

Many social and nutritional factors are important for infectious diseases of poverty. These include the working and living conditions of many of the people who experience food insecurity, poor housing and other hardship and exposure to risk. Health risks that impinge disproportionately on the poor include those for diarrhoeal disease, mosquito-borne infections, Chagas disease, schistosomiasis and sexually transmitted infections. A further example of the links between disease and social deprivation is the occurrence of drug-resistant forms of TB in slums and prisons (*308*).

Several terms in current use, including ecohealth, ecosocial (*284*), biosocial (*309*), eco-bio-social (*310*) and social-ecological (*311*, *312*), attempt to link the ecological, social and biological aspects of health. Recently the term 'syndemics' (*313*) has been linked with One Health (*314*). In common, these concepts call for a systems-based approach. While such calls may seem recent, a partial reaction to risk 'factorology' (*153*), the recognition of inter-connected links, is much older and is at the heart of social medicine (*21*, *315*).

6.7 Success stories

There are several classes of success stories. In the first is the great reduction in infectious diseases of poverty in developed and middle-income countries, including some that are comparatively recent (e.g. the Republic of Korea, Singapore, and China, Province of Taiwan).

The fact that malaria can impede development by lowering economic productivity has long been recognized (*316*), and the same principle no doubt applies to many other conditions that reduce human 'capital', including learning capacity and work output. Such conditions may also cause substantial health costs, though well-targeted and well-implemented health investment can generate a strong economic and health return — this is especially likely if there is a strong public health component to such spending.

The second class relates to targeted interventions in the south, largely generated by money and expertise from the north, but with substantial cooperation from the southern countries. Examples include 1,1,1-trichloro-2,2-di(4-chlorophenyl)ethane (DDT) campaigns against malaria, vaccination campaigns (smallpox, measles, polio) and targeted interventions against dengue fever in Viet Nam using social brigades and copepods for biological control (*317*). The eradication of rinderpest provides another example (*109*). Smallpox has already been eradicated in the wild, and substantial progress has also been made towards eradicating polio. Transmission of guinea worm disease and of onchocerciasis have also been greatly reduced, providing hope for their eventual eradication (*318*).

The prosperity provided by irrigation can generate cultural and other mechanisms that can counter the risk of vector-borne diseases and may lead to their local elimination. This has been termed the 'paddies paradox' (*319*). Other examples of the power of public health over environmental changes are known from the Punjab in pre-independence India, where integrated malaria control was introduced. Integrated malaria control was also successfully introduced in the economically important Zambian copperbelt and sustained for two decades in the 1930s and 1940s (*320*), because the companies involved saw that their economic interests were advanced by providing better health care for their workforce. Vaccination campaigns also fit within this category, as do the largely developed-country-led interventions to reduce SARS in China.

A third class of success story is provided by the large-scale interventions in the south, largely generated by money and expertise from the south. Two good examples are the reduction of the burden of schistosomiasis in China (*321*) and the campaign to decrease Chagas disease in much of Latin America (*322–324*).

The most interesting and the least documented success stories are small-scale interventions in developing countries, largely generated by money and expertise from low-income countries. For example, when Professor Ivo Mueller described how villagers in Papua New Guinea relocated their village to a higher altitude to reduce their perceived increased exposure to mosquito-borne illness caused by warming. Such a story may of course be dismissed as anecdotal, but we include it because it is plausible and hopeful. This category of self-help is also encouraging because it does not rely on perpetually donated funds. The scarcity of such stories may be because they are very rare, but may simply be because no one has reported them.

There is a 'virtuous circle' between improving education, nutrition, knowledge and empowerment (see Box G). This beneficial system of feedback is at the heart of development theory, and can be credited with much of the improvement in public health standards and in life expectancy that occurred in industrializing western Europe (*282, 325*). Such development is likely to lead to many forms of micro-intervention, reducing local disease burdens.

7. Hunger, nutrition, poverty and immunity

In addition to those aspects that we have already discussed, environmental and agricultural change may have other profound influences on the occurrence of infectious diseases of poverty — both directly (via nutrition-related impaired immunity) and indirectly (via exacerbations of hardship, crowding and other manifestations of poverty).

The relationship between infectious diseases of poverty and undernutrition is important, given the very great number of humans who experience deficiencies in both macro- and micronutrients. About half of all child mortality is associated with stunting and other forms of under- and malnutrition (*326*, *327*), while infectious disease is an important risk factor for undernutrition.

'Malnutrition' is often used synonymously for undernutrition, i.e. a deficiency of sufficient protein and/or energy to attain or maintain a normal body height (stunting) or weight (wasting). Foods that supply energy (protein, carbohydrates and fat) are termed macronutrients. However, overnutrition, leading to obesity and chronic disease, is a form of malnutrition. Also to be considered is micronutrient insufficiency — undernutrition of vitamins (especially vitamin A) and/or essential elements, most importantly zinc, iron and iodine (*328*). Today, about one billion people (*329*) experience macronutrient insufficiency, while a far larger number are micronutrient deficient (*330*).

Nevertheless, many people who receive sufficient macro- and micronutrients are quite poorly nourished. Several studies show that, as disposable income falls, people maintain caloric input at the expense of the major micronutrients, shifting to a diet that is comparatively energy dense but nutrient poor (*331*). Although people in this situation may continue to eat the minimum daily requirement of zinc and iron, they are unlikely to consume complex micronutrients, such as flavonoids, and likely to consume excessive amounts of harmful fats. While the long-term health effects of this form of malnutrition are uncertain, they are likely to be harmful (*332*).

Nutritional status and anthropometric measures are also influenced by the health and social-economic circumstances of the individual and population. Children who experience repeated fevers need a greater caloric intake to maintain weight or to grow, as do adults with physically demanding activities. Populations who are chronically parasitized (e.g. by protozoa or helminths) also have a higher caloric demand, as do those who suffer from chronic gut inflammation and impaired absorption. Such circumstances are common in impoverished populations — where undernutrition and illness amplify and reinforce poverty (*333*). Other social factors, including the allostatic load of chronic stress — a measure of the body's cumulative physiological wear and tear in response to chronic environmental demands — conspire to keep poor populations destitute (*334*).

7.1 Links between undernutrition and immunity

The earliest direct link between severe undernutrition (cachexia) and immunological dysfunction was reported in 1810 by Menkel, who observed atrophy of the thymus in malnourished patients (*335*). By the mid-1850s, 'nutritional thymectomy' was becoming a common medical term, although at that time the role of the thymus in immune function was not understood (*336*). Also around this time, Farr recognized the link between starvation and infectious disease, as well as the resulting poverty-related, reinforcing, causal links — including family discord and alcohol abuse (*325, 337*). Undernutrition is a common manifestation of many other diseases, such as HIV/AIDS, and contributes to clinical impairment (*338*).

The association between childhood undernutrition and adult immune function is less clear (*339*). For example, no clear relationship was found between childhood undernutrition and subsequent adult immune function for children conceived during the Dutch 'hunger winter' of 1944–45. However, wider inference from these findings is problematic, since the survivors lived in a developed country with superior health services and low exposure to pathogens, while the period of their undernutrition was much less chronic than that of many people in developing countries.

Analysis of births and deaths in three villages in the Gambia showed that individuals born during the annual 'hungry season' were up to 10 times more likely to die in young adulthood, especially of infectious diseases (*340*). Although this finding has not been duplicated elsewhere (*341*), a more fundamental question remains unanswered: Does antenatal undernutrition 'programme' the fetus to experience impaired immunity in subsequent adulthood? A recent comprehensive review of the effects of both maternal and childhood undernutrition suggests that there is at present insufficient evidence to answer this (*339*).

7.2 Undernutrition and infections: non-immunological links

Acute and chronic undernutrition have other harmful effects upon health, especially in low-income settings, where stunting may also predispose populations to hypertension, cardiovascular disease and type-2 diabetes, especially among individuals who markedly gain weight in later life (*342, 343*). Undernutrition and stunting have profound neurological and mental health effects (*339*), reducing cognitive development, stamina and earning capacity. Taken together, these effects are likely to foster infections and promote other kinds of disadvantage.

As an essential provider of foods that are vital for human well-being and health, agriculture does far more good than harm — even though its capacity to feed the world's growing population is again under serious question (*192*). In the long term, a more concerted and serious international effort to curb population growth, framed in terms of human rights (reproductive choice), might partly address this problem. In the meantime, there is an urgent need to understand the mechanisms that link agricultural activity and food output with the risks of emergence, spread, case severity and fatality of infectious diseases.

7.3 Hunger and the first Millennium Development Goal

Between 1970 and 1996 the number (and percentage) of hungry people declined sharply (from 25% to 14% of the global population). Since then the percentage has remained similar, meaning an increase in absolute terms. Neither of the two recent global targets for hunger reduction can be met. Today approximately one billion people live on less than US$ 1 per day, 162 million of whom live on less than US$ 0.50 (*344*). In 2007 around 800 million people were considered to be food insecure, lacking access to sufficient food to lead healthy, productive lives; however, by early 2009 this had increased to one billion (*329*) (see Figure 9).

Figure 9
Rising number of undernourished

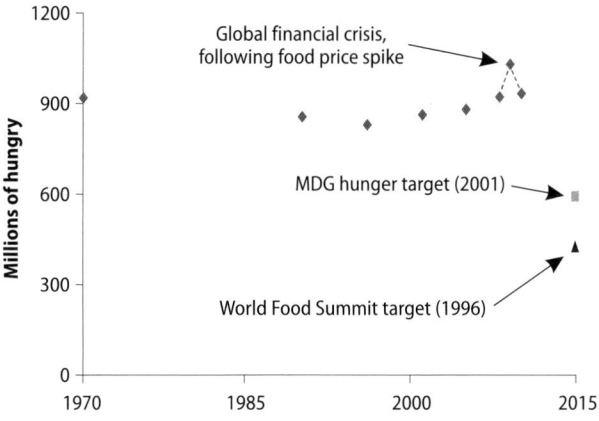

Reducing by half between 1990 and 2015 the proportion of people who live on less than US$ 1 a day and the proportion who suffer from hunger is the first Millennium Development Goal (MDG 1). Poverty reduction and agricultural intensification, expansion and improvement are closely linked in many developing countries. Agriculture is clearly connected also with improved nutrition and health. Fostering positive linkages between the agriculture sector and population health, based on a widened understanding of the nature of this relationship, will enhance progress towards MDG 1, in addition to facilitating attainment of several of the other MDGs.

Until the onset of the global food and financial crises in 2008, some limited progress had been made in reducing poverty, particularly in China, where the proportion of people living on less than US$ 1 a day declined from 29% to 18% between 1990 and 2004 (*345*). Had this rate of reduction continued until 2015, the MDG 1 target would have been met. Nevertheless, even during this period of improvement, in some developing counties, especially in rural parts of Africa and Asia, severe poverty persisted.

Between about 1970 until shortly before the onset of the global financial crisis in 2008, the dominant view held that rapid economic growth and the processes of globalization would cause material gains to 'trickle down' to the extremely poor (*346*). In reality, although the living standards of the middle classes in high-income populations have not improved much over the last 10 years (*347*), the overall impact of globalization and related policies over the last four decades has been high or increasing global inequality (*348*). Disparities in earnings and in income-generating opportunities have increased, and the income of the poorest has fallen further below the absolute poverty line and of national averages.

Three-quarters of the world's ultra-poor live in sub-Saharan Africa. Most chronically impoverished people tend to live in remote rural areas that are located furthest from roads, markets, schools and health services; often face exclusion because of ethnicity, gender or disability; and usually have few assets and less education and access to credit than their urban counterparts (*344*). The chronic undernutrition and other forms of social and environmental deficits that they endure reduce attainment of their full cognitive potential. Also, since they are politically marginalized and unable to escape from poverty unassisted, many are locked in a vicious circle of high fertility, limited birth-spacing and relatively high child mortality rates (*235*).

Poverty and undernutrition are therefore closely linked with each another and with the infectious diseases of poverty — and indeed with many major non-infectious diseases. The poor are therefore vulnerable to infectious diseases of poverty, and their chronic and repeated infections help to 'lock in' further poverty and their undernourished state. Undernutrition thus contributes to this vulnerability. Adverse environmental change — spanning local agricultural practices, regional environmental uses and impacts, and global climate change — adds an extra layer of hardship and health risk to this nexus.

Resolution of this nexus will require complex, sustained, policies with effective instruments. Such policies must be grounded in understanding and engagement across many sectors of government. On its own, the public health sector lacks the skills and resources to undertake such a task. There is therefore need for strategic alliances that strengthen and harmonize relations among all organizations, sectors and institutions concerned with development, environment and social justice. The human rights community is a further important ally,

7.4 Tensions and synergies between agriculture and health

Agriculture and health are linked bidirectionally in many ways. Agriculture is the primary source of livelihood, and hence of food sufficiency and access to basic health care, for most of the world's rural poor. In turn, human health influences agricultural productivity and output. Policy and professional networks that promote agriculture and health should therefore be natural allies. However, there are many barriers that inhibit such cooperation. This is illustrated by tobacco production, a profitable agricultural crop that profoundly harms health. For decades the tobacco industry has tenaciously resisted attempts to reduce this harm (*349*) and the struggle to protect people's health against its product is far from won, especially in developing countries (*350*).

Other examples concern the different interests of the livestock industry, the health sector, and those involved with climate change and local environmental and ecological protection. The continued growth in global population and its dietary aspirations have led to widespread agricultural intensification as well as proliferation of livestock and land use changes that are accelerating production of greenhouse gases and thus climate change (*351, 352*). There is considerable evidence that excessive consumption of animal products is associated with occurrence of heart attacks, strokes, diabetes and, less conclusively, with several forms of cancer (*353, 354*). Foods from animal sources are, however, high in micronutrients and have an important role in the nutrition of vulnerable groups, such as weaning children, women of reproductive age and people with HIV/AIDS. Excessive consumption of sugar (*355*), palm oil, and the subsidies to promote consumption of dairy products under the European Union's Common Agricultural Policy (*356*) further illustrate the conflicting interests of the agriculture and health sectors. Debate has long flared about the safety of pesticides (*357*) and more recently about the environmental and human health costs of genetically modified organisms. Promoters of the use of antibiotics in livestock feed are placing public health at risk, with insufficient evidence of the benefits (*26*).

Apart from tobacco, most of the issues discussed above apply predominantly to the world's over-consumers. Until a few decades ago, high consumers of animal products and sugar lived almost exclusively in high-income countries. Now, more and more live in low- and middle-income countries such as China, India and in South and Central America (*358*). Better distribution systems could stabilize total global livestock production, minimize additional environmental harm and benefit the populations whose consumption of animal foods is either excessive or insufficient (*352*).

7.5 Agriculture and the Millennium Development Goals

The MDG 1 target for hunger reduction is now well beyond reach (see Figure 9). Agriculture is also linked — indirectly or directly — to several other health-related MDG goals: reducing child mortality, improving maternal health, and combating HIV/AIDS, malaria and other diseases (MDG 4, 5, and 6, respectively).

Despite the obvious linkages between agriculture and health, these sectors have often failed to work together in developing joint policies. This may stem partly from a lack of basic awareness of the links in problems and potential solutions and partly from policy conflicts or other obstacles, including those alluded to above. The MDG process also lacks a framework for linking political change, economic policy, and a set of instruments to effectively exploit the potential synergies between agriculture and health.

7.6 Environment, agriculture and health: sectoral cooperation

In order to respond to the issues described in this report, many sectors need to cooperate more closely — cooperation that extends beyond the sectors that are concerned exclusively with infectious diseases to include those that address environmental and social health determinants. An example of this cooperation is the Consultative Group on International Agricultural Research (CGIAR)-associated Expert Group of Agriculture Health Partners, who identified the following research priorities that would benefit from greater intersectoral collaboration, maximize health benefits and lower the environmental and health costs of agricultural activities:

- *Zoonotic diseases and livelihoods.* There is a need to understand the interface between animals and humans and to build capacity to manage the associated risks. This includes carrying out anthropological research on zoonotic diseases, building capacity to quickly identify the pathogen causing an outbreak, and working with producers and market agents to help control livestock diseases. Avian influenza outbreaks provide a unique challenge but also an opportunity to conduct intersectoral research linked to action.
- *Water-associated disease and water management.* Forests are breeding sites for vectors of diseases, aquaculture depends on water, and families need water for cooking, drinking and washing. Agriculture, water and health interact in various ways, some beneficial (e.g. irrigation increases agricultural productivity) and others detrimental (e.g. irrigation water may increase malaria); and the relationships are often bidirectional. Research needed in this area ranges from acquiring new knowledge about the interactions between agriculture, water and health; to developing joint thinking and efforts

to disseminate and apply this knowledge more widely and effectively; to carrying out specific case studies using integrated applicable solutions that can be scaled up. The health aspects of wastewater use in agriculture are becoming increasingly important as such water resources are often available to the peri-urban poor.

Health improvement also depends on transport, energy, science, technology, education, finance and human rights. In many cases, these non-health sectors control substantial resources, including research funds (*20, 359*). Nevertheless, a recent WHO report notes that "In spite of the evidence of an inseparable bidirectional link between health and all facets of human development, galvanizing global attention to the fundamental problem and possible solutions has been slow" (*359*). The need to motivate research funders and governments to focus on the most vulnerable groups in society seems perennial.

7.7 Global action plan

A global action plan would be an appropriate vehicle to bring about the increasingly necessary cooperation between the sectors dealing with infectious diseases, changes to the environment, governance and their associated research disciplines. Ideally, this would be based on extensive consultations with experts and stakeholders from the environment, health and research fields, and would build on the current baseline in these areas.

For example, the Environment and Health Action Plan (2004–2010) in Europe, a follow-up to the European Environment and Health Strategy, is designed to give the European Union a scientifically grounded platform from which to lower the adverse health impacts of environmental change and enhance cooperation between the environment, health and research fields. Three main themes in this plan are to improve knowledge and communication and to review policies.

7.8 Global information systems and databases

There is a need to establish global information systems and databases on the linkages between infectious diseases and environmental changes and conditions. Such initiatives will improve the monitoring of environmental hazards, assessment of the impacts of such hazards on the transmission of infectious diseases and identification of priority areas. Information and monitoring are essential to stimulate appropriate health protection policy, prioritize action, identify and respond to new threats and assess their effectiveness.

Many acute infectious diseases related to environmental change have been tackled successfully (e.g. cholera was last seen in London in the 1850s, with its return being pre-empted by the transformation of sanitation and sewerage systems). However, the capacity for control of infectious disease risks and outbreaks at the global level remains far from secure. Many organizations are potential allies for this task, including WHO, the European Environment Agency, UNEP and the United Nations Development Programme (UNDP). A Global Fund for Zoonoses has also been proposed (*256*).

The links between climate change and some infectious diseases, while challenged by some as simplistic (*28, 360*), are increasingly supported by evidence of such links at the regional level. Climate change is already apparently influencing the distribution of some infectious diseases of poverty (*27, 361*), and although its effect on the burden of other such diseases is currently lower than that of other influences, climate change looms as a major amplifier of such diseases both directly and indirectly.

8. Environment, agriculture and infectious diseases of poverty: selected examples

Environmental changes, including agricultural, influence the epidemiology of infectious diseases independently of nutrition. The impacts are complex and the relationships are typically non-linear. For example, irrigation-associated agriculture can increase breeding opportunities for mosquitoes, and thus the risk of contracting some vector- and waterborne diseases; however, the prosperity provided by irrigation can generate mechanisms that can counter the risks and even lead to local elimination of such diseases (*319*). We focus on three examples of vector-borne diseases to illustrate these complexities, as well as an important waterborne disease, schistosomiasis.

8.1 Vector-borne diseases

8.1.1 Malaria

Malaria, the most important parasitic infection (*362*), is transmitted by many species of *Anopheles* mosquitoes. It has been estimated that *Plasmodium falciparum* caused 200–300 million infections and 1-3 million deaths annually in the 1990s (*363*); however, the increasing use of rapid and accurate diagnostic tests for malaria has revealed that many fevers are misdiagnosed (*31, 364*). Most malaria infection occurs in Africa, but there is intense transmission also in parts of the Western Pacific (*365*). Recently, the comparatively low death toll from malaria in India has been questioned (*366*).

Due to the increased use of insecticide-impregnated bednets, the burden of malaria has declined significantly. Optimists foresee its possible eradication (*367*) but others are more cautious (*367, 368*).

There is evidence that the burden of disease of *P. falciparum* malaria, the most lethal form, may have been fostered by deforestation and slash-and-burn agriculture in Africa about 6000 years ago (*369*). This would probably have provided greater mosquito habitat and a sufficient human population size to maintain transmission.

The presence of livestock can also influence the epidemiology of malaria (*287, 370*) and perhaps that of other mosquito-transmitted infections, such as filariasis. There is also concern that climate change will also influence the epidemiology of malaria, particularly by enabling more intense transmission at higher altitudes (*371*). This issue remains contested (*28, 173, 214*), however, not least because of the paucity of high quality data and of the many known co-factors that alter malarial epidemiology (see Table 5). Lack of cooperation between workers in different disciplines has also hindered progress (*372*). There is nevertheless evidence that malaria is being transmitted at higher altitudes, including in Kenya (*297*), Ethiopia (*373*), Indonesia (*374*) and Papua New Guinea.

P. knowlesi malaria, a newly identified form of the disease, has been described in south-east Asia. Its host species are macaques, the primate that has most successfully adapted to human-dominated landscapes. There is debate as to whether this form of malaria in humans is genuinely new, or simply old but newly identified; a recent paper concludes that it is the latter and that ongoing ecological changes, including deforestation and human population increase, could enable this pathogen to adopt humans as the preferred host (*375*).

Table 5
Factors that alter the dynamics of malaria transmission[a]

Form of change	Example of change	Altered effect[b]
Climate	Temperature, humidity, rainfall[c]	Pathogen life-cycle dynamics, mosquito longevity, activity and distribution
Demographic	Migration, birth rate	Pathogen introduction, immunological status, transmission dynamic alters, especially if malaria is intermittent
Economic	Public health, landscape change, behaviour change	Transmission generally lowered with development
Evolutionary	Pathogen and insecticide resistance	Increases transmission
Landscape	Deforestation, irrigation, urbanization, density of cattle and zoophilia of vectors	Mosquito species dominance, density and activity
Public health	Treatment, insecticides, health education	Number of carriers, disease dynamics altered

[a] While public health and economic development generally reduce malaria transmission, changes in land use, demography and climate may increase or decrease transmission, depending on the local eco-social context.
[b] May be in either direction.
[c] Averages, extremes and distributions.

8.1.2 Dengue fever

Dengue fever, the most important arboviral disease in humans, is transmitted principally by the highly anthropophilic *Aedes (Stegomyia) aegypti* mosquito. A recent review estimated that 2.5 billion people, mostly residents of large and small cities in tropical and sub-tropical countries, are currently at risk of the disease. Today more than 100 tropical countries have endemic dengue virus infections and dengue haemorrhagic fever (DHF) has been documented in more than 60 of them. A total of 50–100 million new dengue infections, accompanied by an estimated 500 000 cases of DHF, are reckoned to occur each year, but surveillance is weak in most countries.

Like malaria, dengue fever has no significant animal reservoir and no vaccine. Unlike malaria, however, there is no curative or prophylactic pharmaceutical treatment. Its epidemiology is clearly related to environmental, social and public health factors (*310*), such as poverty, rapid population growth, urbanization, increased international travel and trade and to the capacity and effectiveness of vector-control programmes, including biological means such as *Mesocyclops* copepods (*317*).

Also like malaria, the epidemiology of dengue fever involves intrinsic population dynamics. Of additional importance are climatic factors that influence deliberate and accidental water storage. Climate change could plausibly play a role in determining the future epidemiology of dengue fever (*376*) but this has been challenged (*377*). Careful analysis, however, does not support the view that climate is irrelevant but rather that numerous cofactors exist (*28, 214*). We are unaware of *any* peer-reviewed literature that has attributed the distribution of dengue fever or malaria solely to climatic factors.

8.1.3 Chagas disease

Chagas disease remains a serious public health problem in many regions of South and Central America due to the wide distribution of host species. It is mainly transmitted by the faecal contamination of skin or conjunctivae that have been broken by the bite of many species of the bug family Reduviidae, principally *Triatoma* spp. and *Rhodnius* spp. All of these vectors originate in and remain well adapted to sylvatic environments, but some are also well adapted to the domestic environment. The domestic vector *Triatoma infestans* has been successfully controlled in Brazil, Chile and Uruguay, though it remains present and thus a threat in sylvatic environments.

As with other zoonoses, Chagas disease is difficult to eradicate since there is always a risk of colonization of human habitat by infected sylvatic and peridomestic bugs. An additional complication is that more than 150 species from 24 families of sylvatic and domestic animals have been found to be infected with *Trypanosoma cruzi*, the etiological agent. The current population of Latin America is around 600 million, half of whom lives in poverty, with about 13% considered to be extremely poor. Estimates indicate that 25% of the Latin American population, predominantly the poor, are at risk of infection with Chagas disease from contact with triatomine insect vectors (*323*). In Colombia, Venezuela, and most countries of Central America, the main vector of Chagas disease is the domiciled insect, *Rhodnius prolixus*. In this region, sylvatic populations of *R. prolixus* have repeatedly re-infested the poor quality rural housing, despite decades of vector control efforts. An analogous situation is occurring further south in the Gran Chaco Region (Argentina, Bolivia, south-western Brazil and Paraguay), where *T. infestans* is re-infesting the human habitat (*378*). Recent reports show that insecticide resistance of *T. infestans* to pyrethroids is a likely contributor to this re-infestation.

8.1.3.1 Biofuel plantations

In the natural cycle, triatomine bugs commonly occur in native palms found in the Amazon and elsewhere in northern South America. Several studies have reported high rates of Chagas infection in these insects, sometimes exceeding 65% (*379*). Over the last 10 years there has been extensive agro-industrial farming of African palm (*Elaeis guineensis*) for oil and biodiesel production in the western plains of Colombia and Venezuela. *R. prolixus* has been able to actively colonize this new ecosystem in Colombia (*380*), which has become the fourth largest global producer of palm oil. Many other extensive plantations are rapidly growing in neighbouring Ecuador, Venezuela and Bolivia. Brazil's ministry of agriculture estimates that the area of cultivation of palm in the Amazon could expand from 60 000 hectares today to more than 6 million hectares in the next 10 years, raising the possibility of reintroduction of infected sylvatic vectors of Chagas disease into this part of Brazil.

Some of this expansion has been driven by a desire to reduce economic dependency on coca plantations (not colonized by triatomes), and has thus probably increased exposure by the poor working populations to triatome vectors (see Table 6). This new example of the 'Columbian exchange' (*100*) has been accompanied by little if any effort to measure the net impact upon human health.

8.1.3.2 Amazon Countries' Initiative for Surveillance and Control of Chagas Disease

Despite increased human disturbance over the last 50 years, the Amazon forest remains an extraordinary repository of biodiversity. While it has long been long inhabited, recent major human impacts include colonization, indiscriminate felling of native forests and mining (see Table 6). In some Latin American countries, forced human migration due to guerrilla activity has also occurred. These impacts can contribute to the emergence and re-emergence of infectious diseases of poverty.

A recently launched vector control effort involving nine Latin American countries is the Initiative of the Amazon Countries for Surveillance and Control of Chagas Disease (AMCHA) (*323*), whose purpose is to evaluate the human impact on the ecosystem and the emergence/re-emergence of Chagas disease.

Table 6
Environmental effects of different drivers and their potential effects on vector-borne diseases in Latin America

Driver	Potential environmental effect	Potential effect on vectors, pathogens and hosts
Climate change	More frequent extreme high temperatures; altered rainfall patterns	Faster development and increased dispersion of vectors and pathogens; possible increases or reduction in vector populations
Colonization of sylvatic environments/ Mining industry/ Increased population growth/ Deforestation	Increased human entry into sylvatic Amazon Region	More human contact with sylvatic vectors More breeding sites Selection of parasite and insect genotypes
Extensive new agro-industrial plantations (*Elaeis guineensis*)	Drastic changes in natural ecosystems	Repercussion still unevaluated
Illicit coca plantations	Drastic changes in natural ecosystems Social conflict	Repercussion partially evaluated
Urbanization	Increased density of human hosts, poorer sanitation, overcrowded human settlements	Increased disease transmission epidemics (dengue fever in Río de Janeiro, yellow fever in Asunción)
El Niño events	Prolonged dry periods in tropical South America, decreased precipitation; increased air temperature	Malaria and dengue outbreaks; new areas of transmission Selection of parasite–reservoir–insect genotypes[a]
Forced human migration	Various, e.g. logging, mining, fires	Urbanization of diseases Increased active and passive transfer of pathogen and vectors

[a] There is a co-evolutionary process between insect vectors and trypanosome transmission. Some triatomine species are able to transmit a particular *Trypanosoma cruzi* population more efficiently. For example, *Triatoma infestans* transmits *T. cruzi II*, while *Rhodnius prolixus* transmits *T. cruzi I* more efficiently. Altered temperatures will favour changed vectors and thus different parasites.

8.1.3.3 Challenges for the future

Like other bloodborne infections, Chagas disease can be transmitted via unscreened blood transfusions, organ transplants, shared needles and poor sterilization practices. More than 100 000 human carriers of Chagas disease are thought to live in the USA and in other countries beyond Central and South America. Bloodborne transmission has been documented in countries where authorities have been less vigilant (*381*). Oral transmission of Chagas disease can also occur via contaminated food following intestinal absorption after passage through the stomach (*382*). Outbreaks transmitted orally have affected up to 100 schoolchildren in Caracas, Venezuela (*383, 384*). In 2009 the Pan American Health Organization (PAHO) published a guide for the surveillance, control and clinical management of acute Chagas disease transmitted by contaminated food (*385*).

Despite these emerging issues, poverty, substandard housing and inadequate public health measures in parts of South America remain the major barriers to controlling Chagas disease. Its association with the biofuel industry is climate sensitive (*386*), and climate change may thus lead to its transmission at higher altitudes than at present.

8.2 Waterborne diseases

Of the many forms of waterborne disease, we focus here on schistosomiasis (bilharzia). Globally, this trematode infection afflicts over 200 million people, of whom at least 80% (*387*) (and perhaps as many as 97%) live in Africa (*388*), with the balance in south-east Asia, the Middle East and Central and South America. The burden of disease of schistosomiasis exceeds that of both TB and malaria (*389*). Like many parasitic infections, there is growing appreciation that schistosomiasis impedes social and economic development. Untreated, it causes substantial morbidity in humans, including anaemia, weakness, ascites, growth retardation and cognitive impairment (*390*). Schistosomiasis due to *Schistosoma haematobium* has disabling and fatal complications including bladder cancer (*390*), while genital forms (*391*) appear to increase the risk of contracting HIV/AIDS (*389*). Currently there are no vaccines for schistosomiasis, and it exemplifies a disease whose control depends on a range of constantly interacting biological, ecological and socioeconomic factors (see Figure 8).

8.2.1 Schistosomiasis in Africa

In Africa, *S. haematobium* and *S. mansoni* are the main parasites that cause schistosomiasis. There are smaller foci of *S. intercalatum* (in the Democratic Republic of the Congo) and *S. guineensis* in West Africa (*392*). The disease is widely distributed in West (*388*), northern and southern Africa, where it causes the death of as many as 300 000 people every year (*389*).

The epidemiology of schistosomiasis in Africa has long been recognized as complex, and subject to numerous social, ecological and climatic factors (*1*). Recent work using climate change models suggests that, although small changes in its distribution are likely, social factors may be more important (*393*).

8.2.2 Schistosomiasis in south-east and east Asia

The main south-east Asian form of schistosomiasis is caused by *S. japonicum* and is endemic mainly in China, Indonesia and the Philippines (*394*). Control is particularly difficult due to its large number of mammalian reservoirs, especially the water buffalo that have traditionally been vital for agriculture. Smaller foci of *S. mekongi* occur in Cambodia and the Lao People's Democratic Republic (*395, 396*). Its reservoir species include pigs and dogs, but apparently not water buffalo (*396*). Transmission of the third species, *S. malayensis*, is restricted to a small region of peninsular Malaysia, where its only known reservoir is rats (*396*).

Considerable progress has been made in reducing the burden of disease from schistosomiasis in this region, especially in China, as described below, and to a lesser extent the Philippines (*394*).

8.2.2.1 Schistosomiasis and climate change in China

Schistosomiasis has been documented in China for over two millennia, but its burden of disease has been greatly lowered by a national control programme that started in the 1950s. At that time the disease was endemic in 12 provinces, infecting an estimated 12 million people and 1 million cattle in areas infested by the intermediate host snail, *Oncomelania hupensis*. However, by late 2003 only 840 000 people and 74 000 cattle were estimated to be infected (*397*).

Regional climate change, especially warming, could have two potential impacts on the frequency and transmission dynamics of *Schistosoma mansoni* in China. First, the current distribution of *O. hupensis* may shift northwards into non-endemic areas. The northern distribution limit of this intermediate host snail is limited to areas where the mean temperature of the coldest month of the year exceeds 0 °C — but this thermal line is likely to shift north (*398*). Warmer temperatures are also predicted to enhance schistosome production by individual snails. The lowest temperature that permits *S. japonicum* development within *O. hupensis* is 15.4 °C; a temperature of 5.8 °C induces hibernation in half the snail population.

A modelling study calculated that by 2050 an additional 748 000 km² of currently non-endemic areas of China would be affected by schistosomiasis, solely based on climate change alone. This area represents 8% of the surface area of China (*398*) and currently supports a population of about 20.7 million people.

9. Research priorities

9.1 Criteria preferences and multi-criteria decision analysis (MCDA) results

The four preferred criteria for research that were identified by the TRG 4 members were:

- Feasibility
- Inter-disciplinarity
- Policy relevance
- Potential to reduce burden of disease

Research priorities 1–10, determined using the multi-criteria decision analysis (MCDA) methodology described in chapter 2, are shown in Table 7, including their scores for the top four criteria. Annex 1 lists all 143 research priorities.

9.2 Relevant research priorities identified by others

Other recent collaborative initiatives have proposed a more integrative approach to determining relevant research priorities.

The first of these pertains to climate change and health, as formulated at the Madrid meeting on Research Priorities for Climate Change Research in October 2008 (http://www.who.int/phe/climate/meeting_madrid/en/).

The second approach deals with agriculture and health. A roundtable on agriculture and health, hosted in 2005 by the International Food Policy Research Institute, and attended by many members of the Consultative Group on International Agriculture Research (CGIAR) and health experts, including some from WHO, led to publication of a special issue of the *Food and Nutrition Bulletin* in 2007 and a summary article in the *Bulletin of the World Health Organization* (*399*). Strategies and priorities for joint international research on agriculture and health resulted also from the meeting: Forging Links between Agriculture and Health, held in Geneva in June 2007 (http://www.cifor.org/publications/pdf_files/research/livelihood/forest_health/pdf2.pdf).

The CGIAR has also developed a major programme on leveraging agriculture for nutrition and health outcomes, including zoonoses, foodborne disease and other infectious diseases of poverty. Of four major themes, two are especially relevant to the present report: prevention and control of agricultural-associated diseases; and integrated agriculture, nutrition, health programmes and policies. The programme also includes a useful list of planned research activities (*400*). Most of these activities are more specific than those we have identified in this report and include risk assessment and prioritization, risk management, communication and capacity strengthening.

Table 7
Top 10 research priorities determined using the multi-criteria decision analysis (MCDA) methodology

Rank	Research priority	Score				Weighted score
		Feasibility	Inter-disciplinarity	Policy relevance	Potential to reduce GBD[a]	
1	Develop integrated preventive public health strategies for infectious diseases of poverty	71	85	86	79	80.5
2	Develop and test novel intersectoral control of neglected tropical diseases	70	90	76	75	78.5
3	Influence funding agencies to support inter-disciplinary approaches to infectious diseases of poverty	74	78	85	74	77.7
4	How to link health, veterinary and wildlife surveillance systems?	69	88	78	64	75.5
5	What groups are most vulnerable to climate change?	75	76	81	67	74.8
6	What are the interactions between agriculture, water use and infectious diseases of poverty?	71	79	76	70	74.3
7	Systems-based research on environmentally induced transmission pathways of vector-borne diseases	66	80	75	73	73.9
8	What is the impact of novel approaches such as community-led total sanitation on helminths?	74	72	75	73	73.4
9	Assess the impacts of water management projects on disease	75	73	77	65	72.5
10	Develop and assess community-based vector-borne disease control models	68	73	76	71	72.0

[a] GBD = global burden of disease.

CGIAR was established in 1971 in response to fears of impending famine; most of its research has focused on increasing on-farm production and productivity and farming systems. More recently, however, it has placed more focus on value chains and markets, and on environmental and health externalities of agriculture. Another of its research programmes focuses on climate change adaptation and mitigation. The CGIAR brings the perspective and experience of agricultural research to bear on those infectious diseases whose epidemiology is influenced by agricultural practices and which can be controlled using agriculture-based interventions (*2*).

A fourth initiative used expert consultation to develop a list of the 100 top questions facing agriculture (*401*). Interestingly, six of these questions focused on pest and disease management, with an emphasis on climate change, environmental sustainability, and vectors — areas that were also included in the TRG 4 list of top-ranked priorities from a biomedical perspective.

9.3 Priorities for policy-makers

This report has reviewed the relationships between the environment (including climate), agriculture and infectious diseases of poverty and identified the key research priorities. It has shown that there is growing recognition of the risks of infectious disease to humans, animals and plants arising from large-scale, human-driven environmental changes, including increasing modifications to the global climate and demand-led agricultural intensification. These major challenges for public health are best met using a systems-based approach to conceptualization and enquiry as well as to determining the relevant research priorities and policy responses. This will require greater understanding of the often complex and non-linear dynamics of ecological relationships and a readiness to widen the scope and interdisciplinary of research thinking, collaboration, and policy development.

The main messages of the report are as follows:

- **As the scale and range of influences on infectious disease emergence and spread become greater, a more collaborative and systems-based approach to prevention and control will yield richer, shared understanding and more effective outcomes.** The various parties involved in reducing the burden of infectious diseases of poverty, improving agriculture and lowering adverse global environmental (and climatic) change have a shared wish to improve long-term human well-being. There is also a growing recognition that the emergence and spread of many infectious diseases reflect fundamental influences from these larger-scale, systemic, environmental and social changes. This invites a more integrative approach to understanding the important contextual (essentially ecological) determinants of infectious disease risks. The use of systems-based concepts and research methods (see also section 1.1.1) potentiates a more collaborative synthesizing approach that can complement, or even replace, the traditional vertically differentiated approach to each 'separate' infectious disease. Further win–win outcomes become more likely if we can avert the inadvertent harm that can result from an exclusive focus on any one of the key factors involved.

- **The burden of existing infectious diseases, particularly as concomitants of poverty, is too high.** The substantial burden of diseases from infections such as malaria, schistosomiasis and hookworm would be considered unacceptably high were the scale, consequences and comparatively modest cost of controlling them better understood. Adverse environmental changes — land use, consequences of intensified food production, human-induced climate change, etc. — compound the task of controlling such diseases. In most regions, however, such changes will probably only modestly increase the total disease burden, assuming that social order and cohesion are sustained.

- **New approaches (concepts, methods and collaborations) are needed.** The burden of infectious diseases of poverty can be lowered in concert with slowing and averting adverse global environmental change. Agricultural intensification and extensification contribute to human well-being, but also cause adverse environmental and climatic changes and thereby amplify the risks of some infectious diseases of poverty. Agricultural failure itself would be even more damaging to human health. Hence, new ways of thinking and acting are needed, including ways to reduce food waste (402), to improve the distribution of food, to minimize the mobilization of novel infectious agents within human populations and, more broadly, to measure 'progress'. A broader coalition of parties interested in reducing poverty and the infectious diseases of poverty is needed.

For at least the next several decades, human economic activities in many parts of the world will tend to disrupt, deplete or otherwise change the natural environment, climate and composition and functioning of ecological systems. This will affect social-ecological structures and systems and may harm human morale on a very large scale. Integrated strategies are needed to deal with these problems and should span the environment, climate, agriculture, social-ecological systems, the microbial world and public health. Inter-disciplinary research and intersectoral action are also crucial.

Despite abundant evidence for the linkages between environmental factors and risks to human health, the contributory causal and mediating processes are not widely understood. Particularly important, but poorly appreciated, is the globally systemic nature of the issues involved, including their relationship with agriculture and infectious diseases of poverty. These involve changes to climate, land use and cover, coastal ecosystems, fisheries and biological diversity, as well as the use of energy and other resources. In turn, these physical aspects interact with social and infrastructural factors, such as population growth and movement, urbanization, trade, social and cultural globalization and transport systems.

Poverty and undernutrition are inextricably linked to many infectious and non-infectious diseases. People who experience poverty and undernutrition are especially vulnerable to infectious diseases, which further reinforce their poverty. Adverse environmental changes intensify this nexus, as do global and regional inequalities in power, wealth, and policy influences. Breaking free from this cycle of entrapment is beyond the capacity of the public health community on its own, and will require strategic alliances and coordinated policy and actions among all those concerned with development, environment and social justice — including human rights — since social and other forms of exclusion are, at core, a human rights issue.

The main cross-cutting conclusions of the report are as follows:

- There have been many recent improvements in global public health, including the development of new vaccines, treatments, and rapid diagnostic tests. Disease transmission continues to be better understood and scientific cooperation remains high (e.g. the increased sharing of information through open-sourced journals). Yet recent global public health gains are increasingly at risk of slowing or even being reversed by the manifest human transformation of the biosphere. This includes global climate change, large-scale soil degradation, biodiversity loss and water and food insecurity. Poverty has remained stubbornly persistent and may worsen because of the rising costs of energy and food, together with a loss of global financial cohesion and confidence. In the worst case, these combined factors may lead to forms of social deterioration and collapse, thereby reducing population immunity and lowering the capacity of health services to provide preventative and curative services.
- There is growing awareness of the potential for emerging infectious diseases to seriously harm health. This concern should be balanced with the recognition that changes in the human (population) host may be equally or more important. Such changes include the growth in size and density of human populations (especially those that are poor, crowded and vulnerable) and the potential reduction in population immunity due to nutritional deficits.

- The potential burdens of poor health and premature death due to a given emerging infectious disease vary considerably. Greater effort should therefore be made to understand the particular characteristics of emerging diseases that determine their disease potential (high, intermediate or low), for both humans and animals.
- Insecticide resistance by disease vectors should be re-instated as a form of disease emergence (see Table 3).
- Some forms of agricultural intensification may be driving increased pathogen virulence, e.g. fast-growing, early-transmitted, and probably more virulent viruses (see Box F).

A strong partnership between science and good governance could, however, meet these infectious disease and nutritional challenges that we are facing. Nevertheless, such a partnership requires people to think beyond their own specialities and boundaries. Today we are facing *systemic* challenges at the local, regional and global levels that are amplified by the increasing connectedness of human populations. There is, therefore, a need for a much broader-based response to the pursuit of continued development and improvement of human lives and health — one that is environmentally sustainable, socio-ecologically sensitive and adaptive to changing conditions.

Itemized and vertically differentiated approaches to infectious disease control will therefore need to be supplemented by and integrated with larger horizontal strategies that ensure environmental sustainability, eco-social sensitivity and adaptive responses (see the main policy recommendations, below). This will require new types and levels of understanding, situation analyses, as well as interdisciplinary research and intersectoral actions to monitor and assess emerging trends and relationships.

The main policy recommendations of the report are as follows:

- Influential people who work in funding bodies, professional societies, and teaching institutions should be encouraged and rewarded for thinking in more systemic, integrative ways, and for adopting and promoting more systemic, horizontal approaches to research and training. Decision-makers in governments and UN bodies should also think and act more systematically (i.e. less vertically), taking a long-term view, based more substantially on scientific findings. If training and professional bodies are thus able to transmit a greater understanding of the complex, interrelated issues involved in the emergence of infectious diseases of poverty, over time this will lead to appropriate changes in other important areas, such as culture and, eventually, governance.

- There are many powerful barriers to systemic thinking, at the conceptual, group and institutional levels. Conceptually, systemic thinking may appear to be boundless, amorphous, difficult and even overwhelming. Group social forces are also an impediment — thinking systemically is difficult to do and to apply in cultures that think in more specialized ways. Group dynamics may, for example, not only fail to reward but even discourage individuals who raise objections to the prevailing view. This applies even more if the objection draws on evidence from other disciplines. For example, an engineer who is aware that a new canal or dam may increase the spread of schistosomiasis, or harm ecological productivity, may feel inhibited from saying so in settings with little perceived understanding or sympathy. And a committee whose mandate is to promote energy security may without hesitation recommend a new coal-fired power station, at the same time discounting or ignoring the resultant climate implications.

- The role of government is to consider and weigh different opinions and to foster policies with the best overall long-term outcomes; however, governments themselves frequently have views and opinions that are not systems based. Furthermore, governments generally act to try to maintain public support. But how can the public support systems-based views if they are rarely taught or supported? Many attempts are also made to shape public and government opinion in ways that favour vested interests, often for policies that erode the ecological and environmental foundations of population health (*403, 404*).

- Systemic thinking, leadership and tenacity are necessary to reduce these barriers. However, the popularity of misinformation augers poorly for global affairs and global health, and again illustrates the importance of policy reform and leadership. Horizontal, integrative thinking offers a chance for synergies to arise through processes of self-organization. But this requires fidelity of information. As Michael Marmot has written: "…people's willingness to take action influences their view of the evidence, rather than evidence influencing their willingness to take action" (*405*). Decosas & Heap conclude: "health ministers are no exception" (*405*).

10. Conclusions

The Thematic Reference Group focused on research needs and challenges concerning interactions between environment, agriculture and infectious diseases of public health importance. Changes in global environment and agricultural systems are among the major overlooked factors in the persistence, emergence and re-emergence of infectious diseases. The changes also interact with trends of economic development, population growth, urbanization, migration and pollution. Climate change and variability add new factors to this conglomerate of driving forces, as do related trends of over- and undernutrition.

The Reference Group identified the following top research priorities for infectious diseases of poverty in relation to environment and agriculture:

1. Develop integrated preventive public health strategies for infectious diseases of poverty.
2. Develop and test novel intersectoral control of neglected tropical diseases.
3. Influence funding agencies to support inter-disciplinary approaches to infectious diseases of poverty.
4. Determine how to link health, veterinary and wildlife surveillance systems.
5. Determine which population groups are most vulnerable to climate change.
6. Determine the interactions between agriculture, water use and infectious diseases of poverty.
7. Apply systems-based research to environmentally induced transmission pathways of vector-borne diseases.
8. Assess the impacts of novel approaches such as community-led total sanitation on helminth infections.
9. Assess the impacts of water management projects on disease.
10. Develop and assess community-based vector-borne disease control models.

Comparing the highest- and lowest-ranked research priorities indicates that the experts' process for eliciting prioritization emphasized options that had a broad rather than a narrow scope; were management-oriented rather than assessment-oriented; incorporated a high level of multidisciplinarity; had an explicit poverty focus; and were focused on impacts rather than outputs.

Application of the systems-based approach described in this report should result in more collaborative, integrated strategies for the prevention and control of infectious diseases, including a more ecologically aware perspective. However, it will require that people modify their way of thinking and that they engage beyond their own specialities — working across sectors, research disciplines and diseases.

Acknowledgements

Special acknowledgement is made by the TRG to Dr Johannes Sommerfeld, Dr Arve Lee Willingham, Dr Ayoade Oduola, Dr Deborah Kioy, Ms Edith Certain, Dr Julie Reza, Dr Margaret Harris, Dr Fabio Zicker and Ms Elisabetta Dessi (all TDR, WHO, Geneva, Switzerland), who were instrumental in creating and managing the TRG, preparing and coordinating the meetings as well as in publishing this report.

The TRG recognizes the valuable support provided by the WHO Representative Office in China, the Chinese Center for Disease Control and Prevention (CDC), under the Chinese Ministry of Health, the Shanghai Municipal Government and the Chongqing CDC/Hubei CDC for assistance in organizing and hosting the stakeholders' consultations.

The TRG was hosted and facilitated by the WHO Country Office in China. Dr Hans Troedsson and Dr Cristobal Tunon (WHO Representative Office in China) were especially helpful and contributed to the stakeholders' consultations. Dr Shao-Hong Lu, Dr Chin-Kei Lee, Dr Ji-Xiang Mao, and Dr Xiao-Dong Zhang, Ms Shan Wuand and Ms Xin-Xin Pang (all of WHO Country Office in China) assisted greatly with organizing and participating in the meetings. Dr Jun Nakagawa (WHO Regional Office for the Western Pacific) also provided technical input and participated in the meetings. Dr Diarmid Campbell-Lendrum (WHO, Geneva) provided additional technical input. Dr Junhao Huang (TDR Career Development Fellow) helped organize and participated in the meetings and was also involved in the preparation of this report.

The TRG extends special thanks to Professor Colin Butler (Associate Professor, National Centre for Epidemiology and Population Health, Australian National University) who assisted greatly in drafting the report. In addition, the Group is also grateful to advisers Dr Delia Grace (International Livestock Research Institute, Nairobi, Kenya) and Professor James Blignaut (Department of Economics, University of Pretoria, Pretoria, South Africa) who contributed to the prioritization of research needs and to drafting the report.

The TRG also acknowledges the valuable contributions made to its work by the stakeholders during the multiple Stakeholders' Consultations.

1. <u>Stakeholders representing governments</u>: Ministry of Health, China: Dr Yang Hao (Director-General, Department of Disease Control); Dr Li-Ying Wang, Dr Lu Ming and Dr Chen Zhao (Director, Deputy Director and Technical Officer, respectively, Division of Schistosomiasis and Other Parasitic Diseases, Department of Disease Control); Dr Yu-Chao Zhao (Director, Sector of Environmental Health, Bureau of Health Supervision); Chinese Center for Disease Control and Prevention (China CDC)): Dr Wei-Zhong Yang, (Vice Director-General), Dr Zheng-Fu Qiang (Director, Department of International

Cooperation), and Dr Zhen Wu (Institute of Environmental Health and Related Product Safety); National Institute of Parasitic Diseases (NIPD), China CDC: Dr Lin-Hua Tang (Director), Dr Jian-Li Kan (Director, Office for Epidemiology), Dr Qi-Yong Liu (Assistant Director), Dr Shi-Zhu Li, Dr Ying-Jun Qiang, Dr Jia-Gang Guo, Dr Wei-Ping Wu, Dr Jia-Xu Chen, Dr Xiao-Hao Wu, Dr Wei Hu, Dr Li-Guang Tian, Dr Tie-Wu Jia, Dr Jia-Wen Yao, Dr Peter Steinmann (visiting scientist from Swiss Tropical and Public Health Institute); Chongqing CDC/Hubei CDC: Mr Ke-Jia Liu (Deputy Director of Health Bureau, Chongqing CDC) and Dr BZ Xiao (Deputy Director of Chongqing CDC); Hainan CDC: Dr Shan-Qing Wang (Deputy Director); Guodong CDC: Dr Yue-Yi Fang (Institute of Parasitic Diseases); Shanghai Municipal Bureau of Health: Dr Shan-Guo Li (Deputy Director of Disease Control Division, Health Department, Shanghai Municipal Government); Shanghai Municipal Center for Disease Control and Prevention: Dr Qi-Zhao Pan (Director) and Dr Chia Li (Chief, Institute of Parasitic Diseases); Royal Norwegian Embassy in Beijing: Dr Werner Christie (Science and Technology Counsellor).

2. <u>Stakeholders representing international and non-governmental organizations</u>: Food and Agriculture Organization of the United Nations (FAO) Representation in China, Democratic People's Republic of Korea and Mongolia: Dr Vincent Martin (Senior Technical Adviser); International Food Policy Research Institute (IFPRI), New Delhi, India office: Dr Suneetha Kadiyala (Platform for Agriculture and Health, Division of Poverty, Health and Nutrition); International Development Research Centre (IDRC), Canada: Dr Dominique Charron (Programme Leader, Ecosystems and Human Health Programme).

3. <u>Stakeholders representing academic institutions</u>: Chinese Academy of Agricultural Sciences (CAAS): Professor Erda Lin (Ex-Director, Institute of Environmental and Sustainable Development in Agriculture), Dr Guang-Zhi Tong and Dr Jiao-Jiao Lin (Director and Deputy Director, respectively, Shanghai Veterinary Research Institute) and Professor Yin-Long Xu; Chinese Academy of Forestry: Professor Xu-Dong Zhang; Zhejiang Academy of Medical Science: Dr Shao-Hong Lu (Deputy Director, Institute of Parasitic Diseases); Fudan University: Professor Qing-Wu Jiang (Dean, School of Public Health), Professor Wei-Dong Qu (Vice-Director, Department of Environmental Health, School of Public Health) and Professor Zhong Yang (School of Life Sciences); University of Ghana: Professor Isabella Akyinbah Quakyi and Professor Chris Gordon; London School of Hygiene & Tropical Medicine: Professor Rosanna Peeling.

Thanks are also extended to Dr Bianca Brijnath (Monash University), Associate Professor Gillian Hall and Dr Haylee Weaver (both of The Australian National University) and Dr Huang Junhao (Zhejiang Agriculture and Forestry University, Hangzhou, China) for their research assistance. We also thank Susan Woldenberg Butler for editorial help.

Acknowledgements

Sincere thanks and appreciation are extended to the peer reviewers, both from within and outside WHO for their comments and contributions on the technical accuracy of this report. Dr Ian G. Neil was responsible for the technical editing of the report.

The activities of the TRG, including the production of this report, were funded by the Special Programme for Research and Training in Tropical Diseases (TDR) and by European Commission, under Agreement PP-AP/2008/160-163.

References

1. Walsh J, Molyneux D, Birley M. Deforestation: effects on vector-borne disease. *Parasitology*, 1993, 106 Suppl:S55-75.
2. Steffen W, Crutzen PJ, McNeill JR. The Anthropocene: are humans now overwhelming the great forces of nature? *Ambio*, 2007, 38:614-21.
3. Bajaj V. As grain piles up, India's poor still go hungry. *New York Times*, 7 June 2012: A1. Available from: http://www.nytimes.com/2012/06/08/business/global/a-failed-food-system-in-india-prompts-an-intense-review.html?_r=1&pagewanted=all
4. Snowden F. Emerging and reemerging diseases: a historical perspective. *Immunological Reviews*, 2008, 225:9-26.
5. Jones KE et al. Global trends in emerging infectious diseases. *Nature*, 2008, 451:990-4.
6. Weiss RA, McMichael AJ. Social and environmental risk factors in the emergence of infectious diseases. *Nature Medicine*, 2004, 10 (Suppl 12):S70-76.
7. Morse SS. Factors in the emergence of infectious diseases. *Emerging Infectious Diseases*, 1995, 1: 7-15.
8. Lederberg J. Infectious disease as an evolutionary paradigm. *Emerging Infectious Diseases*, 1997, 3:417-23.
9. Smolinski M, Hamburg M, Lederberg J. *Microbial threats to health: emergence, detection, and response*. Washington, DC, Institute of Medicine, National Academies Press, 2003.
10. Molyneux DH. Combating the "other diseases" of MDG 6: changing the paradigm to achieve equity and poverty reduction? *Transactions of the Royal Society of Tropical Medicine and Hygiene*, 2008, 102:509-19.
11. Ooi E-E, Gubler DJ. Global spread of epidemic dengue: the influence of environmental change. *Future Virology*, 2009, 4:571-80.
12. Boutayeb A. The double burden of communicable and non-communicable diseases in developing countries. *Transactions of the Royal Society of Tropical Medicine and Hygiene*, 2006, 100:191-9.
13. Chan ED, Iseman MD. Multidrug-resistant and extensively drug-resistant tuberculosis: a review. *Current Opinion in Infectious Diseases*, 2008, 21:587-95.
14. Diamond J. Evolution, consequences and future of plant and animal domestication. *Nature*, 2002, 418(6898):700-7.
15. Ewald P. *Evolution of infectious diseases*. Oxford, Oxford University Press, 1994.
16. Horwitz P, Wilcox BA. Parasites, ecosystems and sustainability: an ecological and complex systems perspective. *International Journal for Parasitology*, 2005; 35: 725-32.
17. Mennerat A et al. Intensive farming: evolutionary implications for parasites and pathogens. *Evolutionary Biology*, 2010, 37(2-3):59-67.
18. Wongsrichanalai C, Meshnick SR. Declining artesunate–mefloquine efficacy against falciparum malaria on the Cambodia–Thailand border. *Emerging Infectious Diseases*, 2008, 14:716-9.
19. Smets BF, Barkay T. Horizontal gene transfer: perspectives at a crossroads of scientific disciplines. *Nature Reviews. Microbiology*, 2005, 3:700-10.
20. Commission on the Social Determinants of Health. *Closing the gap in a generation*. Geneva, World Health Organization, 2008.
21. Eisenberg L. Rudolf Ludwig Karl Virchow, where are you now that we need you? *The American Journal of Medicine*, 1984, 77:524-32.

22. Rayner G, Lang T. *Ecological public health: Reshaping the conditions for good health*. London, Earthscan, 2012.
23. Vilà M et al. Ecological impacts of invasive alien plants: a meta-analysis of their effects on species, communities and ecosystems. *Ecology Letters*, 2011, 14:702-8
24. Mack R, Smith M. Invasive plants as catalysts for the spread of human parasites. *NeoBiota*, 2011, 9:13-29
25. Daszak P, Cunningham AA, Hyatt AD. Emerging infectious diseases of wildlife — threats to biodiversity and human health. *Science*, 2000, 287:443-9.
26. Collignon P et al. The routine use of antibiotics to promote animal growth does little to benefit protein undernutrition in the developing world. *Clinical Infectious Diseases*, 2005, 41:1007-13.
27. Dobson A. Climate variability, global change, immunity, and the dynamics of infectious diseases. *Ecology*, 2009, 90:920-7.
28. Lafferty KD. The ecology of climate change and infectious diseases. *Ecology*, 2009, 90:888-900.
29. Chivian E, Bernstein A, eds. *Sustaining life. How human health depends on biodiversity*. Oxford, Oxford University Press, 2008.
30. *Global report for research on infectious diseases of poverty*. Geneva, World Health Organization, 2012. Available at: www.who.int/tdr/capacity/global_report
31. Gething PW et al. Estimating the number of paediatric fevers associated with malaria infection presenting to Africa's public health sector in 2007. *PLoS Medicine*, 2010, 7(7):e1000301. doi: 10.1371/journal.pmed.1000301
32. Morris S et al. for the Maternal and Child Undernutrition Study Group. Effective international action against undernutrition: Why has it proven so difficult and what can be done to accelerate progress. *Lancet*, 2008, 371:608-21.
33. Swinburn BA et al. The global obesity pandemic: shaped by global drivers and local environments. *Lancet*, 2011, 378:804-14.
34. Pulliam J et al. Agricultural intensification, priming for persistence and the emergence of Nipah virus: a lethal bat-borne zoonosis. *Journal of the Royal Society, Interface*, 2012, 9(66):89-101.
35. Atkinson JA et al. The architecture and effect of participation: a systematic review of community participation for communicable disease control and elimination. Implications for malaria elimination. *Malaria Journal*, 2011, 10:225.
36. Wadsworth Y. *Building in research and evaluation*. Sydney, Allen & Unwin, 2010.
37. Viergever RF et al. A checklist for health research priority setting: nine common themes of good practice. *Health Research Policy and Systems*, 2010, 8:36.
38. Goetghebeur M et al. Combining multicriteria decision analysis, ethics and health technology assessment: applying the EVIDEM decision-making framework to growth hormone for Turner syndrome patients. *Cost Effectiveness and Resource Allocation*, 2010, 8:4.
39. Peacock S et al. Overcoming barriers to priority setting using interdisciplinary methods. *Health Policy*, 2009, 92:124-32.
40. Baltussen R, Niessen L. Priority setting of health interventions: the need for multi-criteria decision analysis. *Cost Effectiveness and Resource Allocation*, 2006, 4:14.
41. McNeill WH. *Plagues and peoples*, 1st ed. Garden City, NY, Anchor Press, 1976.
42. Kapan DD et al. Avian influenza (H5N1) and the evolutionary and social ecology of infectious disease emergence. *EcoHealth*, 2006, 3:187-94.

43. Taylor L, Latham S, Woolhouse M. Risk factors for human disease emergence. *Philosophical Transactions of the Royal Society of London. Series B, Biological Sciences*, 2001, 356:983-39.
44. Froissart R et al. The virulence–transmission trade-off in vector-borne plant viruses: a review of (non-)existing studies. *Philosophical Transactions of the Royal Society of London. Series B, Biological Sciences*, 2010, 365:1907-18.
45. Almeida RPP et al. Spread of an introduced vector-borne banana virus in Hawaii. *Molecular Ecology*, 2009, 18:136-46.
46. Fenwick A. Waterborne infectious diseases — could they be consigned to history? *Science*, 2006, 313: 1077-81.
47. Ostfeld RS, Holt RD. Are predators good for your health? Evaluating evidence for top-down regulation of zoonotic disease reservoirs. *Frontiers in Ecology and the Environment*, 2004, 2:13-20.
48. Marr JS, Cathey JT. New hypothesis for cause of epidemic among native Americans, New England, 1616–1619. *Emerging Infectious Diseases*, 2010, 16: 281-6.
49. Aiello AE et al. Effect of hand hygiene on infectious disease risk in the community setting: a meta-analysis. *American Journal of Public Health*, 2008, 98:1372-81.
50. Stenseth NC et al. Plague dynamics are driven by climate variation. *Proceedings of the National Academy of Science of the United States of America*, 2006, 103: 13110-5.
51. Charrela RN et al. Arenaviruses and hantaviruses: from epidemiology and genomics to antivirals. *Antiviral Research*, 2011, 90: 102-14.
52. Kawaguchi L et al. Seroprevalence of leptospirosis and risk factor analysis in flood-prone rural areas in Lao PDR. *American Journal of Tropical Medicine and Hygiene*, 2008, 78:957-61.
53. ProMed. Leptospirosis, fatal — Sri Lanka (05): northwestern, western, southern, comment. 2011. Available from: http://www.promedmail.org/direct.php?id=20111120.3414
54. Lau CL et al. Climate change, flooding, urbanisation and leptospirosis: fuelling the fire? *Transactions of the Royal Society of Tropical Medicine and Hygiene*, 2009, 104:631-8.
55. McCormick M. Rats, communications, and plague: toward an ecological history. *Journal of Interdisciplinary History*, 2003, 24:1-25.
56. Arjava A. The mystery cloud of 536 CE in the Mediterranean sources. *Dumbarton Oaks Papers*, 2005, 59:73-94.
57. Zhang DD et al. The causality analysis of climate change and large-scale human crisis. *Proceedings of the National Academy of Sciences of the United States of America*, 2011, 108(42):17296-301.
58. McMichael AJ. Insights from past millennia into climatic impacts on human health and survival. *Proceedings of the National Academy of Sciences of the United States of America*, 2012, 109(13):4730-7.
59. Aplin K, Lalsiamliana J. Chronicle and impacts of the 2005-09 mautam in Mizoram. In: Singelton GR et al., eds. *Rodent outbreaks: ecology and impacts*. Los Baños, The Philippines, International Rice Research Institute, 2010:13-48.
60. Clement J et al. Relating increasing hantavirus incidences to the changing climate: the mast connection. *International Journal of Health Geographics*, 2009, 8:1.
61. Klempa B. Hantaviruses and climate change. *Clinical Microbiology and Infection*, 2009, 15:518-23.
62. Pettersson L et al. Outbreak of Puumala virus infection, Sweden. *Emerging Infectious Diseases*, 2008, 14:808-10.
63. Eppig C, Fincher CL, Thornhill R. Parasite prevalence and the worldwide distribution of cognitive ability. *Proceedings of the Royal Society B: Biological Sciences*, 2010, 277(1701):3801-8.

64. Weaver HJ, Hawdon JM, Hoberg EP. Soil transmitted helminthiases: implications of climate change and human behaviour. *Trends in Parasitology*, 2010, 26: 574-81.

65. Greer A, Ng V, Fisman D. Climate change and infectious diseases in North America: the road ahead. *Canadian Medical Association Journal*, 2008, 178:715-22.

66. Hall G, Vally H, Kirk M. Foodborne illnesses: overview. In: Heggenhougen K, Quah S, eds. *International Encyclopedia of Public Health*. San Diego, CA, Academic Press, 2008: 638-53.

67. Ndimubanzi PC et al. A systematic review of the frequency of neurocyticercosis with a focus on people with epilepsy. *PLoS Neglected Tropical Diseases*, 2010, 4:e870.

68. Davey M. Heart of Iowa as as fault line of egg recall. *New York Times*, 27 August 2010: A1. Available from: http://www.nytimes.com/2010/08/27/us/27eggs.html?pagewanted=all

69. Phiri IK, et al. The emergence of *Taenia solium* cysticercosis in Eastern and Southern Africa as a serious agricultural problem and public health risk. *Acta Tropica*, 2003, 87:13-23.

70. Daniel JH et al. Comprehensive assessment of maize aflatoxin levels in eastern Kenya, 2005–2007. *Environmental Health Perspectives*, 2011, 119:1794-9.

71. Stone R. A medical mystery in middle China. *Science*, 2009, 324:1378-81.

72. Kew MC. Interaction between aflatoxin B1 and other risk factors in hepatocarcinogenesis. In: Rai M, Varma A, eds. *Mycotoxins in food, feed and bioweapons*. Springer, Heidelberg, 2010:93-111.

73. Liu Y, Wu F. Global burden of aflatoxin-induced hepatocellular carcinoma: A risk assessment. *Environmental Health Perspectives*, 2010, 118:818-24.

74. Butler CD. Climate change, crop yields, and the future. *United Nations System Standing Committee on Nutrition News (SCN News)*, 2010, 38:18-25.

75. Allan RP. Human influence on rainfall. *Nature*, 2011, 370:344-5.

76. Litaker RW et al. Global distribution of ciguatera causing dinoflagellates in the genus *Gambierdiscus*. *Toxicon*, 2010, 56:711-30.

77. Campora CE et al. Evaluating the risk of ciguatera fish poisoning from reef fish grown at marine aquaculture facilities in Hawai'i. *Journal of the World Aquaculture Society*, 2010, 41:61-70.

78. Llorens J et al. A new unifying hypothesis for lathyrism, konzo and tropical ataxic neuropathy: nitriles are the causative agents. *Food and Chemical Toxicology*, 2011, 49:563-70.

79. Nzwalo H, Cliff J. Konzo: from poverty, cassava, and cyanogen intake to toxico-nutritional neurological disease. *PLoS Neglected Tropical Diseases*, 2011, 5: e1051.

80. Franz E, van Bruggen AHC. Ecology of *E. coli* O157:H7 and *Salmonella enterica* in the primary vegetable production chain. *Critical Reviews in Microbiology*, 2008, 34:143-61.

81. Waltner-Toews D. Food, global environmental change and health: EcoHealth to the rescue? *McGill Journal of Medicine*, 2009, 12:85-9.

82. Srinivasan A et al. Transmission of rabies virus from an organ donor to four transplant recipients. *New England Journal of Medicine*, 2005, 352:1103-11.

83. Aguzzi A, Glatzel M. vCJD tissue distribution and transmission by transfusion — a worst-case scenario coming true? *Lancet*, 2004, 363:411-2.

84. Webster DP, Farrar J, Rowland-Jones S. Progress towards a dengue vaccine. *The Lancet Infectious Diseases*, 2009, 9:678-87.

85. Olotu A et al. Efficacy of RTS,S/AS01E malaria vaccine and exploratory analysis on anti-circumsporozoite antibody titres and protection in children aged 5–17 months in Kenya and Tanzania: a randomised controlled trial. *The Lancet Infectious Diseases*, 2011, 11:102-9.

86. Lightowlers MW. Eradication of *Taenia solium* cysticercosis: a role for vaccination of pigs. *International Journal for Parasitology*, 2010, 40:1183-92.
87. Lienhardt C, Zumla A. BCG: the story continues. *Lancet*, 2005, 366:1414-6.
88. Mas-Coma S, Valero MA, Bargues MD. Effects of climate change on animal and zoonotic helminthiases. *Revue Scientifique et Technique (International Office of Epizootics)*, 2008, 27:443-52.
89. Welsh JS. Contagious cancer. *The Oncologist*, 2011, 16:1-4.
90. *Zoonoses*. Geneva, World Health Organization, 1959 (WHO Technical Report Series No 169).
91. Hart C, Bennett M, Begon M. Zoonoses. *Journal of Epidemiology and Community Health*, 1999, 53:514-5.
92. Hubálek Z. Emerging human infectious diseases: anthroponoses, zoonoses, and sapronoses. *Emerging Infectious Diseases*, 2003, 9:403-4.
93. Gibbons A. American Association of Physical Anthropologists meeting. Tuberculosis jumped from humans to cows, not vice versa. *Science*, 2008, 320:608.
94. Scotch M et al. Human vs. animal outbreaks of the 2009 swine-origin H1N1 influenza A epidemic. *EcoHealth*, 2011, 8:376-80.
95. Smith NH et al. Myths and misconceptions: the origin and evolution of *Mycobacterium tuberculosis*. *Nature Reviews, Microbiology* 2009, 7:537-44.
96. Bhattarai NR et al. Domestic animals and epidemiology of visceral leishmaniasis, Nepal. *Emerging Infectious Diseases*, 2010, 16:231-7.
97. Quaresma PF et al. Wild, synanthropic and domestic hosts of *Leishmania* in an endemic area of cutaneous leishmaniasis in Minas Gerais State, Brazil. *Transactions of the Royal Society of Tropical Medicine and Hygiene*, 2011, 105:579-85.
98. Rotureau B. Are New World leishmaniases becoming anthroponoses? *Medical Hypotheses*, 2006, 67:1235-41.
99. Vasilakis N et al. Fever from the forest: prospects for the continued emergence of sylvatic dengue virus and its impact on public health. *Nature Reviews, Microbiology*, 2011, 9:532-41.
100. Crosby AW. *The Columbian exchange; biological and cultural consequences of 1492*. Westport, CT, Greenwood, 1972.
101. Hotez PJ et al. The neglected tropical diseases of Latin America and the Caribbean: a review of disease burden and distribution and a roadmap for control and elimination. *PLoS Neglected Tropical Diseases*, 2008, 2:e300.
102. Harris JR et al. Multistate outbreak of *Salmonella* infections associated with small turtle exposure, 2007–2008. *Pediatrics*, 2009, 124:1388-94.
103. Chai J-Y, Murrell KD, Lymbery AJ. Fish-borne parasitic zoonoses: Status and issues. *International Journal for Parasitology*, 2005, 35:1233-54.
104. Keiser J, Utzinger J. Emerging foodborne trematodiasis. *Emerging Infectious Diseases*, 2005, 11:1507-14.
105. Rivera-Milla E, Stuermer CAO, Málaga-Trillo E. An evolutionary basis for scrapie disease: identification of a fish prion mRNA. *Trends in Genetics*, 2003, 19:72-5.
106. Rock M et al. Animal–human connections, "one health," and the syndemic approach to prevention. *Social Science & Medicine*, 2009, 68:991-5.
107. Anderson PK et al. Emerging infectious diseases of plants: pathogen pollution, climate change and agrotechnology drivers. *Trends in Ecology and Evolution*, 2004, 19:536-44.

108. Carpenter S, Wilson A, Mellor PS. Culicoides and the emergence of bluetongue virus in northern Europe. *Trends in Microbiology*, 2009, 17:172-8.

109. Morens DM et al. Global rinderpest eradication: lessons learned and why humans should celebrate too. *Journal of Infectious Diseases*, 2011, 204:502-5.

110. Normile D. Driven to extinction. *Science*, 2008,319:1606-9.

111. Gillson L. A 'large infrequent disturbance' in an East African savanna. *African Journal of Ecology*, 2006, 44:458-67.

112. Furuse Y, Suzuki A, Oshitani H. Origin of measles virus: divergence from rinderpest virus between the 11th and 12th centuries. *Virology Journal*, 2010, 7:52.

113. Nava-Ocampo AA et al. Influenza A/H1N1 from a pig perspective. *Canadian Veterinary Journal*, 2009, 50:773-4.

114. Baskin Y. Sea sickness: the upsurge in marine diseases. *BioScience*, 2006, 56:464-9.

115. Smith MD et al. Sustainability and global seafood. *Science*, 2010, 327:784-6.

116. Diaz RJ, Rosenberg R. Spreading dead zones and consequences for marine ecosystems. *Science*, 2008. 321: 926-9.

117. Ferguson HW et al. Jellyfish as vectors of bacterial disease for farmed salmon (*Salmo salar*). *Journal of Veterinary Diagnostic Investigation*, 2010, 22:376-82.

118. Rodger H, Henry L, Mitchell S. Non-infectious gill disorders of marine salmonid fish. *Reviews in Fish Biology and Fisheries*, 2011, 21:423-40.

119. Normile D. Wild birds only partly to blame in spreading H5N1. *Science*, 2006, 312: 1451.

120. LaDeau SL, Kilpatrick AM, Marra PP. West Nile virus emergence and large-scale declines of North American bird populations. *Nature*, 2007, 447:710-3

121. Cunningham A, Daszak P, Rodríguez JP. Pathogen pollution: defining a parasitological threat to biodiversity conservation. *Journal of Parasitology*, 2003, 89(Suppl):S78-S83.

122. Blockstein DE. Lyme disease and the passenger pigeon? *Science*, 1998, 279:1831.

123. Ostfeld RS, Keesing F. The function of biodiversity in the ecology of vector-borne zoonotic diseases. *Canadian Journal of Zoology*, 2000, 78:2061-78.

124. Millennium Ecosystem Assessment, 2005. *Ecosystems and human well-being: biodiversity synthesis*. World Resources Institute, Washington, DC, 2005.

125. Ogden NH et al. Active and passive surveillance and phylogenetic analysis of *Borrelia burgdorferi* elucidate the process of Lyme disease risk emergence in Canada. *Environmental Health Perspectives*, 2010, 118: 909-14.

126. McFarlane R, Becker N, Field H. Investigation of the climatic and environmental context of Hendra virus spillover events 1994–2010. *PLoS One*, 2011, 6: e28374.

127. Luby SP, Gurley ES, Hossain MJ. Transmission of human infection with Nipah virus. *Clinical Infectious Diseases*, 2009, 49: 1743-8.

128. Fenton MB et al. Linking bats to emerging diseases. *Science*, 2006, 311:1084.

129. Bossart GD. Marine mammals as sentinel species for oceans and human health. *Veterinary Pathology*, 2011, 48:676-90.

130. Frick WF et al. An emerging disease causes regional population collapse of a common North American bat species. *Science*, 2010, 329:679-82.

131. Potts SG et al. Global pollinator declines: trends, impacts and drivers. *Trends in Ecology and Evolution*, 2010, 25:345-53.

132. Watanabe ME. Colony collapse disorder: many suspects, no smoking gun. *BioScience*, 2008, 58:384-388.
133. Skerratt LF et al. Spread of chytridiomycosis has caused the rapid global decline and extinction of frogs. *EcoHealth*, 2007, 4:125-34.
134. Wilson EO. *Biophilia*. Cambridge, MA, Harvard University Press, 1984.
135. Stokstad E. Deadly wheat fungus threatens world's breadbaskets. *Science*, 2007, 315:1786-7.
136. Krutmuang P. Climate change effect on insect pest: brown plant hopper and pest management in Thailand. Paper presented at: *Conference on International Research on Food Security, Natural Resource Management and Rural Development, 5–7 October 2011, University of Bonn, Germany*.
137. Zavala J et al. Anthropogenic increase in carbon dioxide compromises plant defense against invasive insects *Proceedings of the National Academy of Science of the United States of America*, 2008, 105: 5129–33.
138. ProMED. Cassava disease threatens food supplies. 2010. Available from: http://www.promedmail.org/direct.php?id=20100525.1735.
139. Gubler DJ. Resurgent vector-borne diseases as a global health problem. *Emerging Infectious Diseases*, 1998, 4:442-50.
140. James CE, Hudson AL, Davey MW. Drug resistance mechanisms in helminths: is it survival of the fittest? *Trends in Parasitology*, 2009, 25:328-35.
141. Bryson V, Demerec M. Bacterial resistance. *The American Journal of Medicine*, 1955, 18:723-37.
142. Parrish CR et al. Cross-species virus transmission and the emergence of new epidemic diseases. *Microbiology and Molecular Biology Review*, 2008, 72:457-70.
143. Kaplan B, Kahn L, Monath T. The brewing storm. *Veterinaria Italiana*, 2009, 45:9-18.
144. Bella H et al. Migrant workers and schistosomiasis in the Gezira, Sudan. *Transactions of the Royal Society of Tropical Medicine and Hygiene*, 1980, 74:36-9.
145. Sleigh AC, Jackson S. Dams, development, and health: a missed opportunity. *Lancet*, 2001, 357:570-1.
146. MacKenzie WR et al. A massive outbreak in Milwaukee of cryptosporidium infection transmitted through the public water supply. *New England Journal of Medicine*, 1994, 331:161-7.
147. Myers S, Patz J. Emerging threats to human health from global environmental change. *Annual Review of Environment and Resources*, 2009, 34:223-52.
148. Oxford JS et al. World War I may have allowed the emergence of "Spanish" influenza. *The Lancet Infectious Diseases*, 2002, 2:111-4.
149. Brown M. Parasites as aetiological agents in chronic diseases. Epidemiological associations and potential mechanisms of pathogenesis. *Parasite Immunology*, 2009, 31:653-5.
150. Carapetis J et al. The global burden of group A streptococcal diseases. *The Lancet Infectious Diseases*, 2005, 5:685-94.
151. Bird A. Perceptions of epigenetics. *Nature*, 2007, 447:396-8.
152. Prüss-Üstün A, Corvalán C. How much disease burden can be prevented by environmental interventions? *Epidemiology*, 2007, 18:167-78.
153. McMichael AJ. Prisoners of the proximate. Loosening the constraints on epidemiology in an age of change. *American Journal of Epidemiology*, 1999, 149: 887-97.
154. Wolfe ND, Dunavan CP, Diamond J. Origins of major human infectious diseases. *Nature*, 2007, 447:279-83.

155. McMichael AJ, Bambrick HJ. The global environment. In: Detels R et al., eds. *Oxford textbook of public health*, 5th ed. Oxford, Oxford University Press, 2009: 220-37.

156. Stillwaggon E. HIV/AIDS in Africa: fertile terrain. *Journal of Development Studies*, 2002, 38:1-22.

157. Le Quere C et al. Trends in the sources and sinks of carbon dioxide. *Nature Geoscience*, 2009, 2:831-6

158. *The global forest resources assessment 2010*. Rome, Food and Agricultural Organization of the United Nations, 2010.

159. Foley JA et al. Solutions for a cultivated planet. *Nature*, 2011, 478:337–42.

160. Webster RG. Wet markets — a continuing source of severe acute respiratory syndrome and influenza? *Lancet*, 2004, 363:234-6.

161. Nayar A. Looking for trouble. *Nature*, 2009, 462:717-9.

162. *Forests and floods. Drowning in fiction or thriving on facts?* Bangkok, Thailand, The Food and Agriculture Organization of the United Nations; Bogar Barat, Indonesia, Center for International Forestry Research, 2005 (RAP Publication 2005/03, Forest Perspectives 2).

163. Hellin J, Haigh M, Marks F. Rainfall characteristics of Hurricane Mitch. *Nature*, 1999, 399:316.

164. Rodriguez-Llanes JM et al. Child malnutrition and recurrent flooding in rural eastern India: a community-based survey. *BMJ Open*, 2011, 2:e000109.

165. Pappas G et al. The globalization of leptospirosis: worldwide incidence trends. *International Journal of Infectious Diseases*, 2008, 12: 351-7.

166. Ko AI et al. Urban epidemic of severe leptospirosis in Brazil. *Lancet*, 1999, 354: 820-5.

167. Bhardwaj P, Kosambiya J, Desai VK. A case control study to explore the risk factors for acquisition of leptospirosis in Surat city, after flood. *Indian Journal of Medical Sciences*, 2008, 62:431-8.

168. Sierra R, Stallings J. The dynamics and social organization of tropical deforestation in northwest Ecuador, 1983-1995. *Human Ecology*, 1998, 26:135-61.

169. Bates SJ et al. Relating diarrheal disease to social networks and the geographic configuration of communities in rural Ecuador. *American Journal of Epidemiology*, 2007, 166:1088-95.

170. Trostle J et al. Raising the level of analysis of food-borne outbreaks: food-sharing networks in rural coastal Ecuador. *Epidemiology*, 2008, 19:384-90.

171. Myers N et al. Biodiversity hotspots for conservation priorities. *Nature*, 2000, 403:853-8.

172. Rival L. The meanings of forest governance in Esmeraldas, Ecuador. *Oxford Development Studies*, 2003, 31:479-501.

173. Gething PW et al. Climate change and the global malaria recession. *Nature*, 2010, 465:342-6.

174. de Castro M et al. Malaria risk on the Amazon frontier. *Proceedings of the National Academy of Science of the United States of America*, 2006, 103(7): 2452-7.

175. Pulliam JRC et al. Epidemic enhancement in partially immune populations. *PLoS ONE*, 2007, e165. doi:10.1371/journal.pone.0000165

176. Plowright RK et al. Reproduction and nutritional stress are risk factors for Hendra virus infection in little red flying foxes (*Pteropus scapulatus*). *Proceedings of the Royal Society B: Biological Sciences*, 2008, 275:861-9.

177. Plowright RK et al. Urban habituation, ecological connectivity and epidemic dampening: the emergence of Hendra virus from flying foxes (*Pteropus* spp.). *Proceedings of the Royal Society B: Biological Sciences*, 2011, 278:3703-12.

178. Fukuda D et al. Bat diversity in the vegetation mosaic around a lowland dipterocarp forest of Borneo. *The Raffles Bulletin of Zoology*, 2009, 57: 213-21.

179. Dobson AP. What links bats to emerging infectious diseases? *Science*, 2005, 310:628.
180. Chua KB, Chua BH, Wang CW. Anthropogenic deforestation, El Nino and the emergence of Nipah virus in Malaysia. *Malaysian Journal of Pathology*, 2002, 24: 15-21.
181. Gurley E, Luby S. Nipah virus transmission in South Asia: Exploring the mysteries, addressing the problems. *Future Virology*, 2011, 6:897-900.
182. Sleigh A, Jackson S. Socio-economic and health impacts of hydropower resettlement projects In: Cleveland C, editor. *Encyclopedia of energy*: Academic Press Reference Series; 2004: 315-23.
183. Syvitski JPM et al. Sinking deltas due to human activities. *Nature Geoscience*, 2009, 2:681-6.
184. Keiser J et al. Effect of irrigation and large dams on the burden of malaria on a global and regional scale. *American Journal of Tropical Medicine and Hygiene*, 2005, 72:392-406.
185. Fenwick A, Cheesmond AK, Amin MA. The role of field irrigation canals in the transmission of *Schistosoma mansoni* in the Gezira Scheme, Sudan. *Bulletin of the World Health Organization*, 1981, 59:777-86.
186. Madsen H et al. Schistosomiasis in Lake Malawi villages. *EcoHealth*, 2011, 8:163-76.
187. Pongsiri MJ et al. Biodiversity loss affects global disease ecology. *BioScience*, 2009, 59:945-54.
188. Steinmann P et al. Schistosomiasis and water resources development: systematic review, meta-analysis, and estimates of people at risk. *The Lancet Infectious Diseases*, 2006, 6:411-25.
189. Srinivasan UT et al. The debt of nations and the distribution of ecological impacts from human activities. *Proceedings of the National Academy of Sciences of the United States of America*, 2008, 105:1768-73.
190. Myers N, Kent J. New consumers: The influence of affluence on the environment. *Proceedings of the National Academy of Science of the United States of America*, 2003, 100:4963-8.
191. Naylor R et al. Losing the links between livestock and land. *Science*, 2005, 310: 1621-2.
192. Godfray HCJ et al. Food security: the challenge of feeding 9 billion people. *Science*, 2010, 327:812-8.
193. Fedoroff NV et al. Radically rethinking agriculture for the 21st century. *Science*, 2010, 327:833-4.
194. Schmidt CW. Swine CAFOs & novel H1N1 flu. Separating facts from fears. *Environmental Health Perspectives*, 2009, 117:A394-401.
195. Hu Z, Shi Z. Investigation of animal reservoir(s) of SARS-CoV. In: Lu Y, Essex M, Roberts B, eds. *Emerging infections in Asia*. New York, NY, Springer, 2008:57-74.
196. Marsh GA, et al. Ebola Reston virus infection of pigs: clinical significance and transmission potential. *Journal of Infectious Diseases* 2011, 204:S804-S809.
197. Butler CD, Weaver HJ. Keep wings from pigs to curb *Reston ebolavirus* (e-letter). *Science*, 2009, 325. Available from: http://www.sciencemag.org/content/325/5937/204.full/reply#sci_el_12539
198. Lebarbenchon C et al. Persistence of highly pathogenic avian influenza viruses in natural ecosystems. *Emerging Infectious Diseases*, 2010, 16:1057-62.
199. Normile D. Rinderpest: driven to extinction. *Science*, 2008,319:1606-9.
200. Li F et al. Finding the real case-fatality rate of H5N1 avian influenza. *Journal of Epidemiology and Community Health*, 2008, 62:555-9.
201. Palese P, Wang TT. H5N1 influenza viruses: Facts, not fear. *Proceedings of the National Academy of Sciences of the United States of America*, 2012, 109(7):2211-3.
202. Rockström J et al. A safe operating space for humanity. *Nature*, 2009, 461:472-5.

203. Intergovermental Panel on Climate Change, 2007. Summary for Policymakers. In: M.L. Parry et al., eds. *Climate Change 2007: Impacts, Adaptation and Vulnerability. Contribution of Working Group II to the Fourth Assessment Report of the Intergovernmental Panel on Climate Change*. Cambridge University Press, Cambridge, England,2007:7-22.

204. Rahmstorf S, Coumou D. Increase of extreme events in a warming world. *Proceedings of the National Academy of Sciences of the United States of America*, 2011, 108(44):17905-9.

205. Trenberth KE. Attribution of climate variations and trends to human influences and natural variability. *Wiley Interdisciplinary Reviews: Climate Change*, 2011, 2:925-30.

206. Romm J. Desertification: the next dust bowl. *Nature*, 2011, 478:450-1.

207. McMichael AJ, Haines A. Global climate change: the potential effects on health. *BMJ*, 1997, 315:805-9.

208. Costello A et al. Managing the health effects of climate change. *Lancet*, 2009, 373: 1693–733.

209. Butler CD, Harley D. Primary, secondary and tertiary effects of the eco-climate crisis: the medical response. *Postgraduate Medical Journal*, 2010, 86:230-4.

210. Reuveny R. Ecomigration and violent conflict: case studies and public policy implications. *Human Ecology*, 2008, 36: 1-13.

211. Mazo J. *Climate conflict*. Abingdon, England, Routledge, 2010.

212. *State of the world's land and water resources for food and agriculture*. Rome, Food and Agriculture Organization of the United Nations, 2011.

213. Hunter PR. Climate change and waterborne and vector-borne disease. *Journal of Applied Microbiology*, 2003, 94:37S-46S.

214. Pascual M, Bouma MJ. Do rising temperatures matter? *Ecology*, 2009, 90:906-12.

215. Meyrowitsch DW et al. Is the current decline in malaria burden in sub-Saharan Africa due to a decrease in vector population? *Malaria Journal*, 2011, 10:188.

216. Garcia-Solache MA, Casadevall A. Global warming will bring new fungal diseases for mammals. *mBio*, 2010, 1:e00061-10.

217. Dherani M et al. Indoor air pollution from unprocessed solid fuel use and pneumonia risk in children aged under five years: a systematic review and meta-analysis. *Bulletin of the World Health Organization*, 2008, 86:390-8.

218. Hall CAS, John W, Day J. Revisiting the limits to growth after peak oil. *American Scientist*, 2009, 97:230-7.

219. McMichael AJ, Butler CD. Promoting global population health while constraining the environmental footprint. *Annual Review of Public Health*, 2011, 32:179-97.

220. Winch P, Stepnitz R. Peak oil and health in low- and middle-income countries: impacts and potential responses. *American Journal of Public Health*, 2011, 101: 1607-14.

221. Howarth RW, Santoro R, Ingraffea A. Methane and the greenhouse-gas footprint of natural gas from shale formations. *Climatic Change*, 2011, 106:679-90.

222. Osborn SG et al. Methane contamination of drinking water accompanying gas-well drilling and hydraulic fracturing. *Proceedings of the National Academy of Sciences of the United States of America*, 2011, 108(20):8172-6.

223. Cordell D, Drangert J-O, White S. The story of phosphorus: global food security and food for thought. *Global Environmental Change*, 2009, 19:292-305.

224. Van Vuuren DP, Bouwman AF, Beusen AHW. Phosphorus demand for the 1970–2100 period: A scenario analysis of resource depletion. *Global Environmental Change*, 2010, 20: 428-39.

225. Bales K. *Disposable people: new slavery in the global economy*. Berkeley, CA, University of California Press, 1999.

226. Coppock DL et al. Capacity building helps pastoral women transform impoverished communities in Ethiopia. *Science*, 2011, 334:1394-8.

227. Sachs JD et al. Ending Africa's poverty trap. *Brookings Papers on Economic Activity*, 2004(1):117-216.

228. Caldwell JC. Routes to low mortality in poor countries. *Population and Development Review*, 1986,12:171-220.

229. Gibson M, Mace R. An energy-saving development initiative increases birth rate and childhood malnutrition in rural Ethiopia. *PLoS Medicine*, 2006, 3(4):e87.

230. Berhane Y. Development projects to improve maternal and child health: assessing the impact. *PLoS Medicine*, 2006, 3(4): e192.

231. Bongaarts J. Human population growth and the demographic transition. *Philosophical Transactions of the Royal Society. Series B, Biological Sciences*, 2009, 364:2985-90.

232. King M, Elliott C. To the point of farce: a Martian view of the Hardinian taboo — the silence that surrounds population control. *BMJ*, 1997, 315:1441-3.

233. Ezeh AC, Bongaarts J, Mberu B. Global population trends and policy options. *The Lancet*, 2012, 380:142-8.

234. Hartmann B. *Reproductive rights and wrongs; the global politics of population control*. Boston, MA, South End Press, 1987.

235. Campbell M et al. Return of the population growth factor. *Science*, 2007, 315:1501-2.

236. Das Gupta M, Bongaarts J, Cleland J. *Population, poverty, and sustainable development : a review of the evidence*. Washington, DC, The World Bank, 2011 (Policy Research Working Paper 5719). Available at: http://www-wds.worldbank.org/external/default/WDSContentServer/IW3P/IB/2011/06/30/000158349_20110630131122/Rendered/PDF/WPS5719.pdf

237. Bryant L et al. Climate change and family planning: least developed countries define the agenda. *Bulletin of the World Health Organization*, 2009, 87:852-7.

238. O'Neill BC et al. Global demographic trends and future carbon emissions. *Proceedings of the National Academy of Sciences of the United States of America*, 2010, 107:17521-6.

239. Rice J, Rice JS. The concentration of disadvantage and the rise of an urban penalty: urban slum prevalence and the social production of health. *International Journal of Health Services*, 2009, 39:749-70.

240. Dye C. Health and urban living. *Science*, 2008, 319:766-9.

241. Harpham T. Urban health in developing countries: What do we know and where do we go? *Health & Place*, 2009, 15:107-16.

242. Firman T et al. Potential climate-change related vulnerabilities in Jakarta: Challenges and current status. *Habitat International*, 2011, 35:372-8.

243. McNeill JR. *Something new under the sun: an environmental history of the twentieth-century world*. New York, NY, WW Norton, 2000.

244. Wilcox BA, Gubler DJ, Pizer HF. Urbanization and the social ecology of emerging infectious diseases. In: Mayer K, Pizer H, eds. *Social ecology of infectious diseases*. Boston, MA, Elsevier/Academic Press, 2007:113-37.

245. Ezzati M et al. Selected major risk factors and global and regional burden of disease. *Lancet*, 2002, 360:1347-60.

246. Barreto FR et al. Spread pattern of the first dengue epidemic in the city of Salvador, Brazil. *BMC Public Health*, 2008, 8:51.
247. Koh BK et al. The 2005 dengue epidemic in Singapore: epidemiology, prevention and control. *Annals of the Academy of Medicine, Singapore*, 2008, 37:538-45.
248. Butler CD. Inequality, global change and the sustainability of civilisation. *Global Change and Human Health*, 2000, 1:156-72.
249. Tricomi E et al. Neural evidence for inequality-averse social preferences. *Nature*, 2010, 463:1089-91.
250. Nowak MA. Five rules for the evolution of cooperation. *Science*, 2006, 314:1560-3.
251. Collier P. *The bottom billion: why the poorest countries are failing and what can be done about it*. Oxford, Oxford University Press, 2007.
252. Butler CD. Environmental change, injustice and sustainability. *Journal of Bioethical Inquiry*, 2008, 5:11-9.
253. Walker B et al. Looming global-scale failures and missing institutions. *Science*, 2009, 325:1345-6.
254. Yamey G. Excluding the poor from accessing biomedical literature: a rights violation that impedes global health. *Health and Human Rights*, 2008, 10:21-42.
255. Perry BD, Grace D, Sones K. Current drivers and future directions of global livestock disease dynamics. *Proceedings of the National Academy of Science of the United States of America*, 2011 (published ahead of print: doi: 10.1073/pnas.1012953108).
256. Tanner M, Zinsstag J. "One Health" — the potential of closer collaboration between human and animal health. *Berliner und Münchener Tierärztliche Wochenschrift*, 2009, 122:410-1.
257. Zinsstag J et al. Potential of cooperation between human and animal health to strengthen health systems. *Lancet*, 2005, 366:2142-5.
258. Zinsstag J et al. Towards a 'One Health' research and application tool box. *Veterinaria Italiana*, 2009, 45:121-33.
259. Waltner-Toews D, Kay J, Lister N. *The ecosystem approach: complexity, uncertainty, and managing for sustainability*. New York, NY, Columbia University Press, 2008.
260. Waltner-Toews D. *One health in an unstable world*. Washington, DC, Institute of Medicine of the National Academies, 2009.
261. Beatty A, Scott K, Tsai P, *rapporteurs. Achieving sustainable global capacity for surveillance and response to emerging diseases of zoonotic origin: Workshop Summary*. Washington, DC, National Academies Press, 2008.
262. Webb J et al. Tools for thoughtful action: the role of ecosystem approaches to health in enhancing public health. *Canadian Journal of Public Health*, 2010, 101: 439-41.
263. Morens DM, Taubenberger JK. Understanding influenza backward. *Journal of the American Medical Association*, 2008, 302:679-80.
264. Gibbs AJ, Armstrong JS, Downie JC. From where did the 2009 'swine-origin' influenza A virus (H1N1) emerge? *Virology Journal*, 2009, 6:207.
265. Charron D, ed. *Ecohealth research in practice: innovative applications of an ecosystem approach to health*. New York, Springer, International Development Research Centre, 2012.
266. Wilcox BA, Aguirre AA, Horwitz P. EcoHealth: connecting ecology, health and sustainability. In: Aguirre AA et al. eds. *Conservation medicine*. Oxford, Oxford University Press, 2012.
267. Bradshaw C, Sodhi N, Brook B. Tropical turmoil — a biodiversity tragedy in progress. *Frontiers in Ecology and the Environment*, 2009, 7:79-87.

268. Bradshaw C et al. Global evidence that deforestation amplifies flood risk and severity in the developing world. *Global Change Biology*, 2007, 13:2379-95.
269. McMichael AJ et al. Global environmental change and health: impacts, inequalities, and the health sector. *BMJ*, 2008, 336:191-4.
270. Walsh PD et al. Catastrophic ape decline in western equatorial Africa. *Nature*, 2003, 422:611-4.
271. Brook B, Sodhi N, Bradshaw C. Synergies among extinction drivers under global change. *Trends in Ecology and Evolution*, 2008, 25:453-60.
272. Gascon C, Williamson GB, da Fonseca GAB. Receding forest edges and vanishing reserves. *Science*, 2000, 288:1356-8.
273. Li R et al. Severe acute respiratory syndrome (SARS) and the GDP. Part I: epidemiology, virology, pathology and general health issues. *British Dental Journal*, 2004, 197:77-80.
274. Peiris J et al. Coronavirus as a possible cause of severe acute respiratory syndrome. *Lancet*, 2003, 361:1319-25.
275. Song H-D et al. Cross-host evolution of severe acute respiratory syndrome coronavirus in palm civet and human. *Proceedings of the National Academy of Sciences of the United States of America*, 2005, 102:2430-5.
276. Guan Y et al. Isolation and characterization of viruses related to the SARS coronavirus from animals in southern China. *Science*, 2003, 302:276-8.
277. Lau SKP et al. Severe acute respiratory syndrome coronavirus-like virus in Chinese horseshoe bats. *Proceedings of the National Academy of Sciences of the United States of America*, 2005, 102:14040-5.
278. Chua KB et al. Nipah virus: a recently emergent deadly paramyxovirus. *Science*, 2000, 288:1432-5.
279. Leroy EM, Gonzalez J-P, Baize S. Ebola and Marburg haemorrhagic fever viruses: major scientific advances, but a relatively minor public health threat for Africa. *Clinical Microbiology and Infection*, 2011, 17:964-76
280. Stevens G, Dias R, Ezzati M. The effects of 3 environmental risks on mortality disparities across Mexican communities. *Proceedings of the National Academy of Sciences of the United States of America*, 2008, 105:16860-5.
281. Bradley DJ. An exploration of chronotones: a concept for understanding the health processes of changing ecosystems. *EcoHealth*, 2004, 1:165-71.
282. Szreter S. Economic growth, disruption, deprivation, disease and death: on the importance of the politics of public health. *Population and Development Review*, 1997, 23:693-728.
283. Hamlin C. *Public health and social justice in the age of Chadwick: Britain 1800-1854*. Cambridge, Cambridge University Press, 1998.
284. Krieger N. Theories for social epidemiology in the 21st century: an ecosocial perspective. *International Journal of Epidemiology*, 2001, 30: 668-77.
285. Vittor AY et al. The effect of deforestation on the human-biting rate of *Anopheles darlingi*, the primary vector of Falciparum malaria in the Peruvian Amazon. *American Journal of Tropical Medicine and Hygiene* 2006, 74:3-11.
286. Yasuoka J, Levins R. Impact of deforestation and agricultural development on Anopheline ecology and malaria epidemiology. *American Journal of Tropical Medicine and Hygiene*, 2007, 76:450-60.
287. Ferguson HM et al. Ecology: a prerequisite for malaria elimination and eradication. *PLoS Medicine*, 2010, 7:e1000303.
288. Ohl C, Tapsell S. Flooding and human health. *BMJ*, 2000, 321:1167-8.

289. Wu X-H et al. Effect of floods on the transmission of schistosomiasis in the Yangtze River valley, People's Republic of China. *Parasitology International*, 2008, 57:271-6.

290. Kondo H et al. Post-flood-infectious diseases in Mozambique. *Prehospital and Disaster Medicine*, 2002, 17:126-33.

291. Ruiz-Moreno D et al. Cholera seasonality in Madras (1901–1940): dual role for rainfall in endemic and epidemic regions. *EcoHealth*, 2006, 4:52–62.

292. Patz J et al. Effects of environmental change on emerging parasitic diseases. *International Journal for Parasitology*, 2000, 30:1395-405.

293. Afrane YA et al. Deforestation and vectorial capacity of *Anopheles gambiae* Giles mosquitoes in malaria transmission, Kenya. *Emerging Infectious Diseases*, 2008, 14:1533-8.

294. Sun Q et al. Response of *Oncomelania* snail distribution on land use in Sichuan, China. *African Journal of Biotechnology*, 2011, 10:13835-40.

295. Jacups SP et al. Predictive indicators for Ross River virus infection in the Darwin area of tropical northern Australia, using long-term mosquito trapping data. *Tropical Medicine & International Health*, 2008, 13:943–52.

296. Chaves LF, Koenraadt CJM. Climate change and highland malaria: fresh air for a hot debate. *The Quarterly Review of Biology*, 2010, 85:27-55.

297. Chen H et al. New records of *Anopheles arabiensis* breeding on the Mount Kenya highlands indicate indigenous malaria transmission. *Malaria Journal*, 2009, 5:17.

298. Sumilo D et al. Climate change cannot explain the upsurge of tick-borne encephalitis in the Baltics. *PLoS ONE*, 2007, 2:e500.

299. Maltezou HC et al. Crimean-Congo hemorrhagic fever in Europe: current situation calls for preparedness. *Eurosurveillance*, 2010, 15: pii=19504.

300. Ogden NH et al. The emergence of Lyme disease in Canada. *Canadian Medical Association Journal*, 2009, 180:1221-4.

301. Semenza JC, Menne B. Climate change and infectious diseases in Europe. *The Lancet Infectious Diseases*, 2009, 9:365-75.

302. Yang G-J, Brook B, Bradshaw C. Predicting the timing and magnitude of tropical mosquito population peaks for maximizing control efficiency. *PLoS Neglected Tropical Diseases*, 2009, 3:e385.

303. Yang G-J et al. Interplay between endogenous and exogenous factors controlling temporal abundance patterns of tropical disease-carrying mosquitoes. *Ecological Applications*, 2008, 18:2028-40.

304. Roberts L. Hitting early, epidemic meningitis ravages Nigeria and Niger. *Science*, 2009, 324:20-1.

305. Mas-Coma S, Valero MA, Bargues MD. Climate change effects on trematodiases, with emphasis on zoonotic fascioliasis and schistosomiasis. *Veterinary Parasitology*, 2009, 163:264-80.

306. Romig T, Thoma D, Weible AK. *Echinococcus multilocularis* — a zoonosis of anthropogenic environments? *Journal of Helminthology*, 2006, 80:207-12.

307. Parkinson AJ, Evengard B. Climate change, its impact on human health in the Arctic and the public health response to threats of emerging infectious diseases. *Global Health Action*, 2009, 2: doi:10.3402/gha.v2i0.2075

308. Koenig R. In South Africa, XDR TB and HIV prove a deadly combination. *Science*, 2008, 319:894-7.

309. Farmer P, Bayona J, Becerra M. Multidrug-resistant tuberculosis and the need for biosocial perspectives. *International Journal Tuberculosis and Lung Diseases*, 2001, 5:885-6.

310. Arunachalam N et al. Eco-bio-social determinants of dengue vector breeding: a multicountry study in urban and periurban Asia. *Bulletin of the World Health Organization*, 2010, 88:173-84.
311. Ostrom E. A general framework for analyzing sustainability of social-ecological systems. *Science*, 2009, 325:419-22.
312. Ostrom E, Janssen M, Anderies J. Going beyond panaceas. *Proceedings of the National Academy of Science of the United States of America*, 2007, 104:15176-8.
313. Singer M, Clair S. Syndemics and public health: reconceptualizing disease in bio-social context. *Medical Anthropology Quarterly*, 2008, 17:423-41.
314. Singer MC. Doorways in nature: syndemics, zoonotics, and public health. A commentary on Rock, Buntain, Hatfield & Hallgrímsson. *Social Science & Medicine*, 2009, 68:991-5.
315. Last J. Fouling and cleansing our nest; human-induced ecological determinants of disease. Supercourse Lecture, 2001, available at: http://www.pitt.edu/~super7/2011-3001/2561.ppt
316. Bynum WF. Mosquitoes bite more than once. *Science*, 2002, 295:47-8.
317. Kay B, Nam VS. New strategy against *Aedes aegypti* in Vietnam. *Lancet*, 2005 365:613-7.
318. Barry M. The tail end of guinea worm — global eradication without a drug or a vaccine. *New England Journal of Medicine*, 2007, 356:2561-4.
319. Ijumba JN, Lindsay SW. Impact of irrigation on malaria in Africa: paddies paradox. *Medical and Veterinary Entomology*, 2001, 15:1-11.
320. Utzinger J et al. The economic payoffs of integrated malaria control in the Zambian copperbelt between 1930 and 1950. *Tropical Medicine and International Health*, 2002, 7:657-77.
321. Utzinger J et al. Conquering schistosomiasis in China: the long march. *Acta Tropica*, 2005, 96:69-96.
322. Dias J, Silveira A, Schofield C. The impact of Chagas disease control in Latin America — a review. *Memórias do Instituto Oswaldo Cruz*, 2002, 97:603-12.
323. Rassi Jr A, Rassi RA, Marin-Neto JA. Chagas disease. *Lancet*, 2010, 375: 1388-402.
324. Monroy C et al. House improvements and community participation in the control of *Triatoma dimidiata* re-infestation in Jutiapa, Guatemala. *Cadernos de Saúde Pública*, 2009, 25 (Suppl 1): S168-S178.
325. Krieger N, Birn A. A vision of social justice as the foundation of public health: commemorating 150 years of the spirit of 1848. *American Journal of Public Health*, 1998, 88:1603-6.
326. Bryce J et al. WHO estimates of the causes of death in children. *Lancet*, 2005, 365: 1147-52.
327. Katona P, Katona J. The interaction between nutrition and infection. *Clinical Infectious Diseases*, 2008, 46:1582-8.
328. Black RE et al. Maternal and child undernutrition: global and regional exposures and health consequences. *Lancet*, 2008, 371:243-60.
329. *The state of food insecurity in the world*. Rome, Food and Agricultural Organization of the United Nations, 2009.
330. Black R. Micronutrient deficiency — an underlying cause of morbidity and mortality. *Bulletin of the World Health Organization*, 2003, 81:79.
331. Hoddinott J et al. Effect of a nutrition intervention during early childhood on economic productivity in Guatemalan adults. *Lancet*, 2008, 371:411-6.
332. Cordain L et al. Origins and evolution of the Western diet: health implications for the 21st century. *American Journal of Clinical Nutrition*, 2005, 81:50-4.
333. Humphrey JH. Child undernutrition, tropical enteropathy, toilets, and handwashing. *Lancet*, 2009, 374:1032-5.

334. Evans G, Schamberg M. Childhood poverty, chronic stress, and adult working memory. *Proceedings of the National Academy of Science of the United States of America*, 2009, 106:6545-9.
335. Beisel W. History of nutritional immunology: introduction and overview. *Journal of Nutrition*, 1992, 122:591-6.
336. Turk JL. Paul Ehrlich — the dawn of immunology. *Journal of the Royal Society of Medicine*, 1994, 87:314-5.
337. Hamlin C. Could you starve to death in England in 1839? The Chadwick–Farr controversy and the loss of the "social" in public health. *American Journal of Public Health*, 1995, 85:856-66.
338. Keusch GT. The history of nutrition: malnutrition, infection and immunity. *Journal of Nutrition*, 2003, 133:336S-40S.
339. Victora CG et al. Maternal and child undernutrition 2: Maternal and child undernutrition: consequences for adult health and human capital. *Lancet*, 2008, 371:340-57.
340. Moore SE et al. Season of birth predicts mortality in rural Gambia. *Nature*, 1997, 388:434.
341. Moore SE. Commentary: patterns in mortality governed by the seasons. *International Journal of Epidemiology*, 2006, 35:435-7.
342. Kuh D et al. Offspring birth weight, gestational age and maternal characteristics in relation to glucose status at age 53 years: evidence from a national birth cohort. *Diabetic Medicine*, 2008, 25:530-5.
343. Barker DJ et al. Growth and chronic disease: findings in the Helsinki Birth Cohort. *Annals of Human Biology*, 2009, 36:445-58.
344. Ahmed A et al. *The world's most deprived: characteristics and causes of extreme poverty and hunger*. Washington, DC, International Food Policy Research Institute; 2007.
345. Chen S, Ravallion M. *Absolute poverty measures for the developing world, 1981–2004*. Washington, DC, World Bank, 2007.
346. Dollar D. Is globalization good for your health? *Bulletin of the World Health Organization*, 2001, 79:827-33.
347. Krugman P. The Big Zero. *New York Times*, 28 December 2009: A27. Available from: http://www.nytimes.com/2009/12/28/opinion/28krugman.html
348. Wade RH. On the causes of increasing world poverty and inequality, or why the Matthew effect prevails. *New Political Economy*, 2004, 9:163-88.
349. Drope J, Chapman S. Tobacco industry efforts at discrediting scientific knowledge of environmental tobacco smoke: a review of internal industry documents. *Journal of Epidemiology and Community Health*, 2001, 55:588-94.
350. China's unhealthy relations with big tobacco. *Lancet*, 2011, 377:180.
351. Steinfeld H et al. *Livestock's long shadow*. Rome, Food and Agricultural Organization of the United Nations, 2006.
352. McMichael AJ et al. Food, agriculture, energy, climate change and health. *Lancet*, 2007, 370:1253-63.
353. Popkin BM. Reducing meat consumption has multiple benefits for the world's health. *Archives of Internal Medicine*, 2009, 169:543-5.
354. Sinha R et al. Meat intake and mortality. A prospective study of over half a million people. *Archives of Internal Medicine*, 2009, 169:562-71.
355. Lang T, Heasman M. *Food wars: the global battle for mouths, minds and markets*. London, Earthscan, 2004.

356. Lloyd-Williams F et al. Estimating the cardiovascular mortality burden attributable to the European Common Agricultural Policy on dietary saturated fats. *Bulletin of the World Health Organization*, 2008, 86:535–41.

357. Pimentel D et al. Environmental and economic costs of pesticide use. *BioScience*, 1992, 42:750-60.

358. Garrett JL, Ruel MT. *Stunted child-overweight mother pairs: an emerging policy concern?* Washington, DC, International Food Policy Research Institute, 2003 (FCND Discussion Paper Briefs No. 148).

359. *Research and development coordination and financing: Report of the Expert Working Group.* Geneva, World Health Organization, 2010.

360. Randolph SE. Perspectives on climate change impacts on infectious diseases. *Ecology*, 2009, 90:927-31.

361. Wilson K. Climate change and the spread of infectious ideas. *Ecology*, 2009, 90: 901-2.

362. Greenwood BM et al. Malaria. *Lancet*, 2005, 365:1487-98.

363. Breman J. The ears of the hippopotamus: manifestations, determinants, and estimates of the malaria burden. *American Journal of Tropical Medicine and Hygiene*, 2001, 64(Suppl): 1-11.

364. Bell D, Peeling RW. Evaluation of rapid diagnostic tests: malaria. *Nature Reviews. Microbiology*, 2009, 44:S34-S8.

365. Snow RW. Malaria risk: Estimating clinical episodes of malaria (reply). *Nature*, 2005, 437: E4-E5. doi:10.1038/nature04180

366. Dhingra N et al. Adult and child malaria mortality in India: a nationally representative mortality survey. *Lancet*, 2010, 376:1768-74.

367. Shah NK. Assessing strategy and equity in the elimination of malaria. *PLoS Medicine*, 2010, 7:e1000312.

368. Gosling R, Chandramohan D. Tackling malaria today: beware resurgence of malaria where incidence has fallen. *BMJ*, 2008, 337:a1592.

369. Pennisi E. Malaria's beginnings: on the heels of hoes? *Science*, 2001, 293:416-7.

370. Mutero CM et al. A transdisciplinary perspective on the links between malaria and agroecosystems in Kenya. *Acta Tropica*, 2004, 89:171-86.

371. Martens P et al. Climate change and future populations at risk of malaria. *Global Environmental Change*, 1999, 9:S89-S107.

372. Brisbois BW, Harris S. Climate change, vector-borne disease and interdisciplinarity research: social science perspectives on an environment and health controversy. *EcoHealth*, 2010, 7:425-38.

373. ProMed. Ethiopia sounds alarm over new malaria prone areas. Afrik.com, 2010. Available from: http://www.promedmail.org/direct.php?id=20100501.1414

374. ProMed. Malaria — Indonesia: (Papua) suspected, request for information. 2010. Available from: http://www.promedmail.org/direct.php?id=20100611.1965

375. Lee K-S et al. *Plasmodium knowlesi*: reservoir hosts and tracking the emergence in humans and macaques. *PLoS Pathogens*, 2011, 7:e1002015.

376. Hales S et al. Potential effect of population and climate changes on global distribution of dengue fever: an empirical model. *Lancet*, 2002, 360:830-4.

377. Reiter P. Climate change and mosquito-borne disease. *Environmental Health Perspectives*, 2001, 109(Suppl 1):141-61.

378. Gürtler R. Sustainability of vector control strategies in the Gran Chaco Region: current challenges and possible approaches. *Memórias do Instituto Oswaldo Cruz*, 2009, 104(Suppl I):52-9.

379. Guhl F, al. Actualización de la distribución geográfica y ecoepidemiología de la fauna de triatominos (Reduviidae: Triatominae) en Colombia [Updated geographical distribution and ecoepidemiology of the triatomine fauna (Reduviidae: Triatominae) in Colombia]. *Biomedica*, 2007, 27(Suppl 1):143-62 (in Spanish).

380. Fitzpatrick S et al. Molecular genetics reveal that silvatic *Rhodnius prolixus* do colonise rural houses. *PLoS Neglected Tropical Diseseas*, 2008, 2:e210.

381. Igreja RP. Chagas disease 100 years after its discovery. *Lancet*, 2009, 373:1340.

382. Nóbrega AA et al. Oral transmission of Chagas disease by consumption of açaí palm fruit, Brazil. *Emerging Infectious Diseases*, 2009, 15:653-5.

383. Alarcón de Noya B et al. Large urban outbreak of orally acquired acute Chagas disease at a school in Caracas, Venezuela. *Journal of Infectious Diseases*, 2010, 201:1308-15.

384. Miles MA. Orally acquired Chagas disease: lessons from an urban school outbreak. *Journal of Infectious Diseases*, 2010, 201:1282-4.

385. Pan American Health Organization. Doença de Chagas: Guia para vigilancia, prevencão, controle e manejo clínico da doença de chagas aguda transmitida por alimentos. PAHO/HS/CD/539.09. 2009. Available at: http://www.saude.rs.gov.br/upload/1335550488_Guia%20Para%20 Vigil%C3%A2ncia,%20Preven%C3%A7%C3%A3o,%20Controle%20e%20Manejo.pdf (accessed 22 July 2012).

386. Bustamante DM et al. Environmental determinants of the distribution of Chagas disease vectors in south-eastern Guatemala. *Geospatial Health*, 2007, 2:199-211.

387. Simoonga C et al. Remote sensing, geographical information system and spatial analysis for schistosomiasis epidemiology and ecology in Africa. *Parasitology*. 2009, 136:1683-93.

388. Schur N et al. Geostatistical model-based estimates of schistosomiasis prevalence among individuals aged ≤20 years in West Africa. *PLoS Neglected Tropical Diseases*, 2011, 5:e1194.

389. Hotez PJ, Fenwick A. Schistosomiasis in Africa: an emerging tragedy in our new global health decade. *PLoS Neglected Tropical Diseases*, 2008, 3:e485.

390. King CH, Dangerfield-Cha M. The unacknowledged impact of chronic schistosomiasis. *Chronic Illness*, 2008, 4:65-79.

391. Rollinson D. A wake up call for urinary schistosomiasis: reconciling research effort with public health importance. *Parasitology*, 2009, 136:1593-610.

392. Moné H et al. Human schistosomiasis in the Economic Community of West African states: epidemiology and control. *Advances in Parasitology*, 2010, 71:33-91.

393. Stensgaard A-S et al. Large-scale determinants of intestinal schistosomiasis and intermediate host snail distribution across Africa: Does climate matter? *Acta Tropica*, available ahead of print at: http://dx.doi.org/10.1016/j.actatropica.2011.11.010 (accessed 1April 2012)

394. Bergquist R, Tanner M. Controlling schistosomiasis in Southeast Asia: a tale of two countries. *Advances in Parasitology*, 2010, 72:109-44.

395. Attwood SW, Fatih FA, Upatham ES. DNA-sequence variation among *Schistosoma mekongi* populations and related taxa; phylogeography and the current distribution of Asian schistosomiasis. *PLoS Neglected Tropical Diseases*, 2008, 2:e200.

396. Conlan JV et al. A review of parasitic zoonoses in a changing Southeast Asia. *Veterinary Parasitology*, 2011, 182:22-40.

397. Zhou XN et al. The public health significance and control of schistosomiasis in China — then and now. *Acta Tropica*, 2005, 96:97-105.

398. Zhou X-N et al. Potential impact of climate change on Schistosomiasis transmission in China. *American Journal of Tropical Medicine and Hygiene*, 2008, 78:188-94.

399. Hawkes C, Ruel M. The links between agriculture and health: an intersectoral opportunity to improve the health and livelihoods of the poor. *Bulletin of the World Health Organization*, 2006, 84:984-90.

400. CGIAR. Agriculture for Improved Nutrition and Health. CGIAR Research Program 4 (http://www.cgiarfund.org/cgiarfund/sites/cgiarfund.org/files/Documents/PDF/crp_4_Proposal.pdf): International Food Policy Research Institute; 2011.

401. Pretty J et al. The top 100 questions of importance to the future of global agriculture. *International Journal of Agricultural Sustainability*, 2010, 8:219-36.

402. Parfitt J, Barthel M, MacNaughton S. Food waste within food supply chains: quantification and potential for change to 2050. *Philosophical Transactions of the Royal Society of London. Series B, Biological Sciences*, 2010,365:3065-81.

403. Oreskes N, Conway EM. *Merchants of doubt: How a handful of scientists obscured the truth on issues from tobacco smoke to global warming*. New York, Bloomsbury Press, 2010.

404. Pearse G. *High & dry: John Howard, climate change and the selling of Australia's future*. Melbourne, Penguin/Viking, 2007.

405. Decosas J, Heap S. Where is the charcoal? *Lancet*, 2004, 364:1896-8.

Annex 1

Research priorities ranked 1-143, determined using the multi-criteria decision analysis (MCDA) methodology

Rank	Priority
1	Develop integrated preventive public health strategies for infectious diseases of poverty.
2	Develop and test novel intersectoral control of neglected tropical diseases.
3	Influence funding agencies to support interdisciplinary approaches to infectious diseases of poverty.
4	Determine how to link health, veterinary and wildlife surveillance systems.
5	Determine which groups are most vulnerable to climate change.
6	Determine the interactions between agriculture, water use and infectious diseases of poverty.
7	Systems-based research on environmentally induced transmission pathways of vector-borne diseases.
8	Assess the impact of novel approaches such as community-led total sanitation on helminths.
9	Assess the impacts of water management projects on disease.
10	Develop and assess community-based vector-borne disease control models.
11	What are the environmental health risks for infectious diseases of poverty?
12	Remote sensing to improve prediction of vector-borne disease outbreaks.
13	How to improve surveillance for climate sensitive diseases?
14	What are the priority agriculture-related infectious diseases of poverty and why?
15	Research intersectoral interventions to control infectious diseases of poverty.
16	Assess the impact of climate change on food production, biodiversity and nutrition.
17	Strengthen community infectious diseases of poverty initiatives to improve sustainability.
18	Understand longer-term impacts of climate change on human health.

(continued)

Rank	Priority
19	What is the relation between wildlife–livestock interaction and disease emergence?
20	Assess the impact of climate change on malaria and the impact of development projects on malaria vectors.
21	What are the health risks of climate variability?
22	Develop data observatories for agriculture- and environment-related infectious diseases of poverty.
23	Contribute to environment, social and health impact assessment for large development projects.
24	Which diseases are climate sensitive and how might they alter under climate change?
25	What is the fraction of disease burden attributable to the environment?
26	What are the implications of emerging infectious disease for global health security?
27	What is the impact of expanding agriculture on zoonotic emerging diseases?
28	Develop models to assess climate change impacts on infectious disease transmission.
29	What adaptive strategies for climate change do or could vulnerable groups use?
30	What fraction of environment-attributable disease is amenable to reduction?
31	What has been learned about environment- and agriculture-related infectious diseases of poverty and what are the gaps?
32	How do water, agriculture and health interact?
33	How to promote health impact assessment in agriculture projects and appropriate response?
34	Map vulnerable populations and their malaria burden.
35	Establish level and type of malnutrition (under and over) that causes immune impairment and disease.
36	Develop new technologies and methodologies to better manage infectious diseases of poverty.
37	What are the relations between early disturbance of ecosystems, food production and emerging disease?

(continued)

Rank	Priority
38	Develop early warning for epidemics.
39	What are the broad impacts of industrial agriculture plantations (including illegal plantations)?
40	Assess health implications of decisions made in other sectors.
41	Model the relation between climate change impacts on agriculture and human health.
42	What are the health implications of adaptive strategies to climate change?
43	Monitor the impacts of climate change on vector-borne zoonoses.
44	Assess effectiveness and cost-benefit of interventions for mitigating health impacts of climate change
45	What is the relation between gender, poverty and the health and livelihood benefits of agriculture?
46	Assess new and existing methods for dengue control (e.g. predatory fish, water covers).
47	Investigate how to protect the poor from wildlife zoonoses.
48	Understanding the interactions between diet, nutrition and health.
49	Understand transmission pathways of zoonoses impacting on livelihoods.
50	How can research on agriculture, water and health reduce animal and foodborne disease?
51	What are the lessons from previous attempts to integrate environment, agriculture and infectious diseases of poverty?
52	How to understand factors that hinder or promote intersectoral integration?
53	What are the drivers for disease emergence from animals?
54	What is the impact of environment change and vector behaviour?
55	Assess the impact of agriculture development programmes on food safety.
56	Assess how to understand relations between HIV/AIDS, agriculture and nutrition in order to design integrated programmes.
57	Improve public health workers' understanding of global environmental change.
58	Explore how to develop integrated agriculture, health and nutrition programmes.

(continued)

Rank	Priority
59	Understand the human/animal interface to identify and control zoonoses.
60	What are the health aspects of wastewater use?
61	Develop decision support tools for assessing health impacts of climate change.
62	Evaluate effectiveness interventions in other sectors (agriculture, industry, etc.) compared with interventions at individual/household level to reduce climate change-driven health effects.
63	What are the health impacts of animal faeces in water?
64	What are the health risks of urban environments?
65	How to understand, monitor and value ecosystem disease regulating services?
66	Develop sustainable dengue vector control that works despite climate change.
67	Assess the disease burden and other non-health impacts of agriculture associated diseases.
68	How to control malaria among immigrant/mobile populations?
69	Develop improved vulnerability and adaptation assessments.
70	Develop pattern recognition analysis to generate evidence of climate change impacts on disease.
71	What are the health risks of forced human migration?
72	How to understand and monitor drivers of foodborne diseases?
73	Assess effects of industry waste in poor countries.
74	Participatory approaches to engage general population in understanding climate change and disease.
75	Develop pattern recognition analysis to understand change in infectious disease.
76	How to promote climate change adaptation policies?
77	What is the contribution of climate change to health burden?
78	Undertake environmental and climate impact studies on malaria.
79	At what threshold does climate change affect specific health problems?
80	How to strengthen public health systems to adapt to climate change?
81	What is the role of different actors in addressing agriculture-associated disease?

(continued)

Rank	Priority
82	Assess the relation between economy changes and food security in low-income areas.
83	What is the impact of urban agriculture on infectious diseases of poverty?
84	What is the impact of climate change on migration?
85	What is the relation between social networks and vulnerability to foodborne disease?
86	What are interactions between climate change and health determinants?
87	Assess climate change health adaptation plans in high-risk countries.
88	Assess the impact of climate change on diarrhoeal disease of farm workers and foodborne disease.
89	Assess how to disseminate climate change health research to policy- and decision-makers.
90	Understand the context in which interdisciplinary approaches are developing and generating evidence of impacts.
91	What are the factors that make some new diseases high-impact?
92	What are mechanisms for climate change affecting pathogens and vectors?
93	What is impact of wildlife trade (for food/pets) on emerging infectious disease?
94	Develop tools to distinguish ecological changes.
95	Develop methods to better attribute foodborne diseases to different causes.
96	Assess how to monitor diarrhoea in poor communities to identify changes due to climate change.
97	Assess how to reduce risk of animal disease (zoonotic and non-zoonotic).
98	Use influence diagrams to understand complex systems in which altered disease risk occur.
99	Assess the links between expanding markets, agriculture and infectious diseases of poverty.
100	Assess economic returns to health.
101	Develop innovations to improve home water storage.
102	How does population movement due to climate change affect health services?
103	How to improve post-harvest technologies for food security?

(continued)

Rank	Priority
104	Assess livelihood impacts of response to zoonotic diseases (e.g. avian 'flu).
105	Improve integration of climate change mitigation, adaptation and health.
106	How to remediate contaminated water?
107	How to address under- and over-nutrition in poor countries?
108	How to reduce the impact of climate change on schistosomiasis?
109	What is the public health worker's understanding of climate change and infectious diseases of poverty?
110	Develop and test health education and mass media propaganda.
111	Improve the assessment of the impacts of mining/timber extraction/oil extraction on infectious diseases of poverty.
112	What is the additional impact of climate change on infectious diseases of poverty above other drivers?
113	Assess the effectiveness of short-term interventions to mitigate the health effects of climate change.
114	Investigate how to encourage cooperatives for agriculture production.
115	Develop diagnostic tools for surveillance of environment-induced infectious diseases of poverty.
116	Assess how to manage the benefits and risks of animal source foods.
117	Use multi-modelling to assess the impact of health policies on non-health sectors.
118	Improve uptake of risk mitigation by poor producers and consumers.
119	Develop and adapt innovative strategies for personal protection.
120	Assess the relationship between economic changes and food safety in low-income areas.
121	How to use value chains to improve nutrition?
122	Understanding social, environment and economic drivers of disease emergence from bushmeat.
123	Research into agriculture and urban waste management.
124	How can relevant risks to health and livelihood be reduced without reducing agricultural productivity?

(continued)

Rank	Priority
125	What is vulnerable groups' understanding of climate change and infectious diseases of poverty?
126	How to raise awareness on new diseases in vulnerable communities to support reporting through non-governmental organisations?
127	Monitor impacts of climate change on ticks and tick-borne disease.
128	What are the drivers of bat population change and hence increased risk of zoonoses?
129	How to promote compost use in Africa?
130	How to remediate environments after primary industry activities?
131	Identify nutritious/high-yield plants for wider use.
132	How does agricultural use of drugs affect antibiotic resistance?
133	What is the impact of climate change on employment and access to health?
134	How to detect insecticide residues?
135	How to ensure food security where cash cropping is practised?
136	What is the role of secondary vectors in malaria transmission?
137	What is the relation between agrochemicals, other factors and cholera?
138	Research on 'efficient microbes' and waste management/vector control.
139	How to recycle plastic waste?
140	Assess effectiveness and impact of genetically modified mosquitoes on malaria.
141	Assess the impacts of resource constraints on mental health and infectious diseases of poverty.
142	Assess the safety of genetically modified food.
143	Develop a strategy to eliminate monkey malaria (*Plasmodium knowlesi*).

Appendices

Appendix 1

Membership of Thematic Reference Group on Environment, Agriculture and Infectious Diseases of Poverty (TRG4)

	Names	Country	Expertise	Gender
CO-CHAIRS	Prof Anthony McMichael	Australia	Epidemiology and Population health	M
	Prof Xiao-Nong Zhou	China	Clinical parasitologist (Infectious diseases)	M
MEMBERS	Prof Corey Bradshaw	Australia	Ecological modelling and Environment	M
	Dr Stuart Gillespie	Switzerland	Agriculture and Food policy	M
	Prof Colin Butler	Australia	Environmental science and Public health	M
	Prof Suad M. Sulaiman	Sudan	Health and Environment	F
	Prof James A. Trostle	USA	Anthropology	M
	Prof Jürg Ützinger	Switzerland	Public health and Epidemiology	M
	Prof Bruce A. Wilcox	USA	Global health and Disease ecology	M
	Dr Guojing Yang	China	Infectious diseases	F
	Prof Felipe Guhl (2008–2009)	Colombia	Microbiology and parasitology	M
	Dr A. Lee Willingham (2008–2009)	USA/ Denmark	Food safety and Zoonoses	M

Appendix 2
Disease-specific and thematic reference groups (DRGs/TRGs) of The Think Tank for infectious diseases of poverty and host countries

Reference group		Host institution and country
DRG1	Malaria	WHO Regional Office for Africa, Congo
DRG2	Tuberculosis, leprosy and Buruli ulcer	WHO Country Office, Philippines
DRG3	Chagas disease, human African trypanosomiasis and leishmaniasis	WHO Country Offices, Sudan and Brazil
DRG4	Helminth infections	African Programme for Onchocerciasis Control (APOC), Burkina Faso
DRG5	Dengue and other emerging viral diseases of public health importance	WHO Country Office, Cuba
DRG6	Zoonoses and marginalized infectious diseases of poverty	WHO Regional Office for the Eastern Mediterranean, Egypt
TRG1	Social sciences and gender	WHO Country Office, Ghana
TRG2	Innovation and technology platforms for health interventions in infectious diseases of poverty	WHO Country Office, Thailand
TRG3	Health systems and implementation research	WHO Country Office, Nigeria
TRG4	Environment, agriculture and infectious diseases of poverty	WHO Country Office, China

Appendix 3
Composition of the TDR Think Tank

Professor Pedro Alonso, Director and Research Professor, Barcelona Centre for International Health Research (CRESIB), Barcelona, Spain

Professor Rose Leke, Head, Department of Microbiology, Immunology, Hematology and Infectious Diseases, Faculty of Medicine and Biomedical Sciences, University of Yaoundé, Yaoundé, Cameroon

Dr Joel Breman, Senior Scientific Adviser, Fogarty International Center, Division of International Epidemiology & Population Studies, National Institutes of Health, Bethesda, MD, USA

Professor Graham Brown, Foundation Director, Nossal Institute for Global Health, University of Melbourne, Carlton, Victoria, Australia

Dr Chetan Chitnis, Principal Investigator, International Centre for Genetic Engineering and Biotechnology (ICGEB), New Delhi, India

Professor Alan Cowman, Researcher, Walter and Eliza Hall Institute of Medical Research, Parkville, Victoria, Australia

Professor Abdoulaye Djimdé, Research Scientist, Chief of Laboratory, Malaria Research and Training Center (MRTC), University of Bamako, and Malian EDCTP Senior Fellow, Bamako, Mali

Dr Sócrates Herrera Valencia, Director, Caucaseco Scientific Research Center (SRC), Instituto de Inmunología del Valle, Malaria Vaccine & Drug Development Centre, Universidad del Valle, Cali, Colombia

Professor Marcelo Jacobs-Lorena, Johns Hopkins School of Public Health, Department of Molecular Microbiology and Immunology, Malaria Research Institute, Baltimore, MD, USA

Dr Ramanan Laxminarayan, Director, Center for Disease Dynamics, Economics and Policy (CDDEP), Washington, DC, USA

Professor Rosanna Peeling, Chair of Diagnostics Research, London School of Hygiene & Tropical Medicine, Department of Infectious and Tropical Diseases, Clinical Research Unit, London, England

Professor Akintunde Sowunmi, University College Hospital, Malaria Research Laboratories, Institute of Advanced Medical Research and Training (IAMRAT), Ibadan, Nigeria

Dr Sarah Volkman, Senior Research Scientist, Harvard School of Public Health, Department of Immunology and Infectious Diseases, Boston, MA, USA

Dr Tim Wells, Chief Scientific Officer, Medicines for Malaria Venture (MMV), Geneva, Switzerland

Professor Gavin Churchyard, Chief Executive Officer, Aurum Institute, Johannesburg, South Africa

Professor Charles Yu, Medical Director and Vice President for Medical Services, De La Salle Health Sciences Institute, Vice-Chancellor's Office for Mission, Cavite, Philippines

Dr Madhukar Pai, Assistant Professor, McGill University, Department of Epidemiology, Biostatistics & Occupational Health, Montreal, Quebec, Canada

Dr Ann M. Ginsberg, Senior Adviser, Global Alliance for TB Drug Development, New York, NY, USA

Dr Jintana Ngamvithayapong-Yanai, President, TB/HIV Research Foundation, Chiang Rai, Thailand

Professor Laura C. Rodrigues, Head, Department of Epidemiology and Population Health, London School of Hygiene & Tropical Medicine, London, England

Professor Martien Borgdorff, Head, Cluster Infectious Diseases, Municipal Health Service of Amsterdam and Professor of Epidemiology, University of Amsterdam, Amsterdam, Netherlands

Professor Biao Xu, Director of Tuberculosis Research Center, Professor of Epidemiology and Deputy Chair, Department of Epidemiology, School of Public Health, Fudan University, Shanghai, China

Dr Francis Adatu Engwau, Programme Manager, National Tuberculosis/Leprosy Programme, Kampala, Uganda

Dr Anthony David Harries, Senior Adviser, Director, Department of Research, London School of Hygiene & Tropical Medicine, London, England

Dr Timothy Paul Stinear, Head of Research Group NHMRC, R. Douglas Wright Research Fellow, Department of Microbiology and Immunology, University of Melbourne, Parkville, Victoria, Australia

Dr Helen Ayles, Director, ZAMBART Project, London School of Hygiene & Tropical Medicine, ZAMBART, Ridgeway Campus, University of Zambia, Lusaka, Zambia

Professor Diana Lockwood, Department of Infectious and Tropical Diseases, London School of Hygiene & Tropical Medicine, London, England

Professor Ken Stuart, President Emeritus & Founder, Seattle Biomedical Research Institute, Seattle, WA, USA

Professor Maowia M. Mukhtar, Institute of Endemic Diseases, Department of Molecular Biology, University of Khartoum, Khartoum, Sudan

Professor Bianca Zingales, Instituto de Quimica, Universidade de São Paulo, São Paulo, Brazil

Professor Marleen Boelaert, Head, Department of Public Health, Institut de Médecine Tropical, Epidemiology & Disease Control Unit, Department of Public Health, Antwerp, Belgium

Ms Marianela Castillo-Riquelme, Departamento de Economía de la Salud, DIPLAS, Subsecretaría de Salud Publica, Ministerio de Salud de Chile, Santiago, Chile

Professor Mike J. Lehane, Professor of Molecular Entomology and Parasitology, Liverpool School of Tropical Medicine, Liverpool, England

Professor Pascal Lutumba, Institut National de Recherche Bio-Médicale, Kinshasa University, Kinshasa, Democratic Republic of the Congo

Dr Enock Matovu, Senior Lecturer, Faculty of Veterinary Medicine, Makerere University, Department of Veterinary, Parasitology and Microbiology, Kampala, Uganda

Dr David Sacks, Head, Intracellular Parasite Biology Section, National Institutes of Health, National Institute of Allergy and Infectious Diseases, Laboratory of Parasitic Diseases, Bethesda, MD, USA

Dr Sergio Alejandro Sosa-Estani, Head, Service of Epidemiology, Instituto de Efectividad Clinica y Sanitaria, Buenos Aires, Argentina

Dr Shyam Sundar, Department of Medicine, Institute of Medical Sciences, Banaras Hindu University, Varanasi, India

Professor Rick L. Tarleton, Distinguished Research Professor, Center for Tropical & Emerging Global Diseases, Coverdell Center for Biomedical Research, University of Georgia, Athens, GA, USA

Professor Alon Warburg, Professor of Vector Biology and Parasitology, The Kuvin Center for the Study of Infectious and Tropical Diseases, Faculty of Medicine, Hebrew University, Ein Kerem, Israel

Dr Sara Lustigman, Head, Laboratory of Molecular Parasitology, Lindsley F. Kimball Research Institute, New York Blood Center, New York, NY, USA

Dr Boakye Boatin, Noguchi Memorial Institute for Medical Research, University of Ghana, Legon, Accra, Ghana

Dr Guojing Yang, Vice Head, Department of Schistosomiasis Control, Jiangsu Institute of Parasitic Diseases, Wuxi, China

Dr Rashida M.D.R. Barakat, High Institute of Public Health, Alexandria University, Alexandria, Egypt

Dr Maria Gloria Basanez, Professor of Neglected Tropical Diseases, Department of Infectious Disease Epidemiology, Faculty of Medicine, Imperial College, London, England

Dr Kwablah Awadzi, Onchocerciasis Chemotherapy Research Centre, Hohoe Hospital, Hohoe, Ghana

Professor Banchob Sripa, Division of Experimental Pathology, Department of Pathology, Faculty of Medicine, Khon Kaen University, Khon Kaen, Thailand

Professor Warwick Grant, Head of Genetics, School of Molecular Sciences, Genetic Department, La Trobe University, Bundoora, Victoria, Australia

Professor Roger K. Prichard, Professor of Biotechnology, Institute of Parasitology, McGill University, Ste Anne de Bellevue, Quebec, Canada

Professor Hector Hugo Garcia, Department of Microbiology and Cysticercosis Unit, Instituto de Ciencias Neurológicas, Universidad Peruana Cayetano Heredia, Lima, Peru

Dr James McCarthy, Group Leader, Clinical Tropical Medicine, Queensland Institute of Medical Research, University of Queensland, Herston, Queensland, Australia

Professor Kouakou Eliezer N'Goran, Professeur de Biologie, Laboratoire de Zoologie et de Biologie Animale, Université de Cocody, Abidjan, Côte d'Ivoire

Dr Andréa Gazzinelli, School of Nursing, Federal University of Minas Gerais, Belo Horizonte, MG, Brazil

Dr Jeremy Farrar, Director, Oxford University Clinical Research Unit in Viet Nam, The Hospital for Tropical Diseases, Ho Chi Minh City, Viet Nam

Professor Maria Guzman, Head, Virology Department, Instituto de Medicina Tropical "Pedro Kouri", Havana, Cuba

Dr Natarajan Arunachalam, Senior Grade Deputy Director, Centre for Research in Medical Entomology, Indian Council of Medical Research, Madurai, India

Dr Duane Gubler, Professor, Director, Asia-Pacific Institute of Tropical Medicine and Infectious Diseases, John A Burns School of Medicine, University of Hawaii, Honolulu, HI, USA

Dr Sirirpen Kalayanarooj, Queen Sirikit National Institute of Child Health, Bangkok, Thailand

Dr Linda Lloyd, Director, Center for Research, The Institute for Palliative Medicine at San Diego Hospice, San Diego, CA, USA

Dr Lucy Chai See Lum, Associate Professor of Paediatrics, Department of Paediatrics, Faculty of Medicine, University of Malaya Medical Centre, Kuala Lumpur, Malaysia

Dr Amadou Sall, Chef de l'Unité des Arbovirus et Virus des Fièvres hémorragiques, Insitut Pasteur de Dakar, Arboviruses Unit/WHO Collaborating Centre and Conference Centre, Dakar, Senegal

Dr Eric Martinez Torres, Instituto de Medicina Tropical Pedro Kouri, Havana, Cuba

Dr Philip J. McCall, Vector Group, Liverpool School of Tropical Medicine, Liverpool, England

Professor Derek Cummings, Assistant Professor, Department of Epidemiology, Bloomberg School of Public Health, Johns Hopkins University, Baltimore, MD, USA

Dr Hongjie Yu, Deputy Director, Professor, Office for Disease Control and Emergency Response, Chinese Center for Disease Control and Prevention, Beijing, China

Professor David Molyneux, Senior Professorial Fellow, Liverpool School of Tropical Medicine, Liverpool, England

Dr Zuhair Hallaj, Senior Consultant on Communicable Diseases, WHO Regional Office for the Eastern Mediterranean, Cairo, Egypt

Dr Gerald T. Keusch, Professor of International Health and of Medicine, Boston University, Boston, MA, USA

Dr Pilar Ramos-Jimenez, Philippine NGO Council on Population, Health and Welfare, Pasay City, Philippines

Professor Donald Peter McManus, National Health and Medical Research Council of Australia, Senior Principal Research Fellow, Head of Molecular Parasitology Laboratory, Queensland Institute of Medical Research, Brisbane, Queensland, Australia

Dr Eduardo Gotuzzo, Director, Instituto de Medicina Tropical "Alexander von Homboldt", Universidad Peruana Cayetano Heredia, Lima, Peru

Dr Kamal Kar, Chairman, CLTS Foundation, Calcutta, India

Dr Ana Sanchez, Associate Professor, Department of Community Health Sciences, Brock University, St. Catharines, Ontario, Canada

Dr Amadou Garba, Director, Réseau International Schistosomose, Environnement, Aménagement et Lutte (RISEAL), Niamey, Niger

Dr Helena Ngowi, Department of Veterinary Medicine and Public Health, Sokoine University of Agriculture, Mongoro, United Republic of Tanzania

Dr Sarah Cleaveland, Reader, Division of Ecology and Evolutionary Biology, University of Glasgow, Glasgow, Scotland

Dr Hélène Carabin, University of Oklahoma, Oklahoma Health Sciences Center, Oklahoma City, OK, USA

Professor Barbara McPake, Director and Professor, Institute for International Health and Development, Queen Margaret University, Edinburgh, Scotland

Dr Margaret Gyapong, Director, Dodowa Health Research Centre, Ghana Health Service, Dodowa, Ghana

Professor Juan Arroyo Laguna, Profesor Principal del Departamento Académico de Salud y Ciencias Sociales, FASPA-UPCH, Universidad Pruana Cayetano Heredia, Lima, Peru

Professor Sarah Atkinson, Reader, Department of Geography, University of Durham, Science Laboratories, Durham, England

Professor Rama Baru, Professor, Centre of Social Medicine and Community Health, Jawaharlal Nehru University, New Delhi, India

Professor Otto Nzapfurundi Chabikuli, Regional Technical Director, Africa Region with Family Health International (FHI360), Pretoria, South Africa

Professor Kalinga Tudor Silva, Senior Professor, Faculty of Arts, University of Peradeniya and Executive Director, International Centre for Ethnic Studies, Kandy, Sri Lanka

Professor Charles Hongoro, Research Director, Policy Analysis Unit, Human Sciences Research Council, Pretoria, South Africa

Professor Mario Mosquera-Vasquez, Associate Professor, Departamento de Comunicación Social, Universidad del Norte, Barranquilla, Colombia

Professor Chuma Jane Mumbi, Research Fellow, Kenya Medical Research Institute, Wellcome Trust Research Programme, Kilifi, Kenya

Professor Helle Samuelsen, Head, Department of Anthropology, University of Copenhagen, Copenhagen, Denmark

Professor Sally Theobald, Liverpool School of Tropical Medicine, Liverpool, England

Professor Mitchell Weiss, Professor and Head of the Department of Public Health and Epidemiology, Swiss Tropical Institute, Basel, Switzerland

Professor Yongyuth Yuthavong, Senior Researcher, National Centre for Genetic Engineering and Biotechnology (BIOTEC), Bangkok, Thailand

Professor Simon Croft, Professor of Parasitology, Department of Infectious and Tropical Diseases, London School of Hygiene & Tropical Medicine, London, England

Professor Rama Baru, Professor, Centre of Social Medicine and Community Health, Jawaharlal Nehru University, New Delhi, India

Professor Sanaa Botros, Manager of Training and Consultation Unit, Theodor Bilharz Research Institute, Imbaba, Giza, Egypt

Dr Mary Jane Cardosa, Director, Institute of Health and Community Medicine, University Malaysia Sarawak, Kota, Malaysia

Professor Simon Efange, Professor of Chemistry, University of Buea, Buea, Cameroon

Dr Vish Nene, Director of Biotechnology Thematic Group, International Livestock Research Institute, Nairobi, Kenya

Dr Antonio Oliveira-Dos-Santos, Medical Affairs Director, Genzyme, Rio de Janeiro, Brazil

Professor Paul Reider, Department of Chemistry, Princeton University, Princeton, NJ, USA

Dr Giorgio Roscigno, Former Chief Executive Officer, Foundation for Innovative New Diagnostics, Budé, Geneva, Switzerland

Professor Anthony So, Director, Terry Stanford Institute of Public Health Policy, Duke University, Durham, NC, USA

Professor Ming-Wei Wang, Director, The National Centre for Drug Screening, Shanghai, China

Dr Miguel Angel González-Block, Executive Director, Centre for Health Systems Research, National Institute of Public Health, Cuernavaca, Mexico

Professor Olayiwola Akinsonwon Erinosho, Executive Secretary at Health Reform Foundation of Nigeria (HERFON), Abuja, Nigeria

Dr Charles Collins, Honorary Senior Research Fellow, University of Birmingham, Birmingham, England

Dr Dyna Arhin, Associate Consultant, Public Health Action Support Team (PHAST), Faculty of Medicine, Imperial College London, England

Dr Abbas Bhuiya, Senior Social Scientist, Head, Poverty and Health Programme and Social and Behavioural Sciences Unit, Public Health Sciences Division, ICDDR,B, Mohakhali, Dhaka, Bangladesh

Dr Celia Maria de Almeida, Senior Researcher and Professor in Health Policy and Health Systems Organization, Health Administration and Planning Department, Escola Nacional de Saúde Pública-ENSP/Fiocruz, Rio de Janeiro, Brazil

Professor Barun Kanjilal, Professor, Indian Institute of Health Management Research, Jaipur, India

Dr Joseph Kasonde, Executive Director, Zambia Forum for Health Research, Lusaka, Zambia

Dr Dorothée Kinde-Gazard, Minister of Health, The National AIDS Control Programme (PNLS), Cotonou, Benin

Dr Samuel Wanji, Research Foundation for Tropical Diseases and the Environment, Buea, Cameroon

Professor Anthony McMichael, Professor, National Centre for Epidemiology and Population Health, Australian National University, Canberra, ACT, Australia

Professor Xiao-Nong Zhou, Director, National Institute of Parasitic Disease, China Center for Disease Control, Shanghai, China

Professor Corey Bradshaw, Director of Ecological Modelling, The Environment Institute and School of Earth & Environmental Sciences, University of Adelaide, Adelaide, Western Australia, Australia

Professor Colin D. Butler, National Centre for Epidemiology and Population Health, The Australian National University, Canberra, Australia

Dr Stuart Gillespie, Director, RENEWAL, Coordinator, Agriculture and Health Research Platform, International Food Policy Research Institute (IFPRI), c/o UNAIDS, Geneva, Switzerland

Professor Felipe Guhl , Centro de Investigaciones en Microbiología y Parasitología Tropical, Facultad de Ciencias, Universidad de los Andes, Bogotá, Colombia (2008–2009)

Dr Suad M. Sulaiman, Health & Environment Adviser, Khartoum, Sudan

Professor James A. Trostle, Professor of Anthropology, Anthropology Department, Trinity College, Hartford, CT, USA

Dr Jürg Ützinger, Assistant Professor, Department of Public Health and Epidemiology, Swiss Tropical Institute, Basel, Switzerland

Professor Bruce A. Wilcox, Tropical Disease Research Laboratory, Department of Pathology, Faculty of Medicine, Khon Kaen University, Khon Kaen, Thailand

Dr Arve Lee Willingham, Deputy Director, WHO/FAO Collaborating Centre for Research and Training for Neglected and Other Parasitic Zoonoses, University of Copenhagen, Denmark (2008–2009)

Dr Guojing Yang, Assistant Professor (Principal Investigator), Department of Schistosomiasis Control, Jiangsu Institute of Parasitic Diseases, Jiangsu Province, China

Appendix 4

Distribution of the Think Tank leadership (*co-Chairs*)

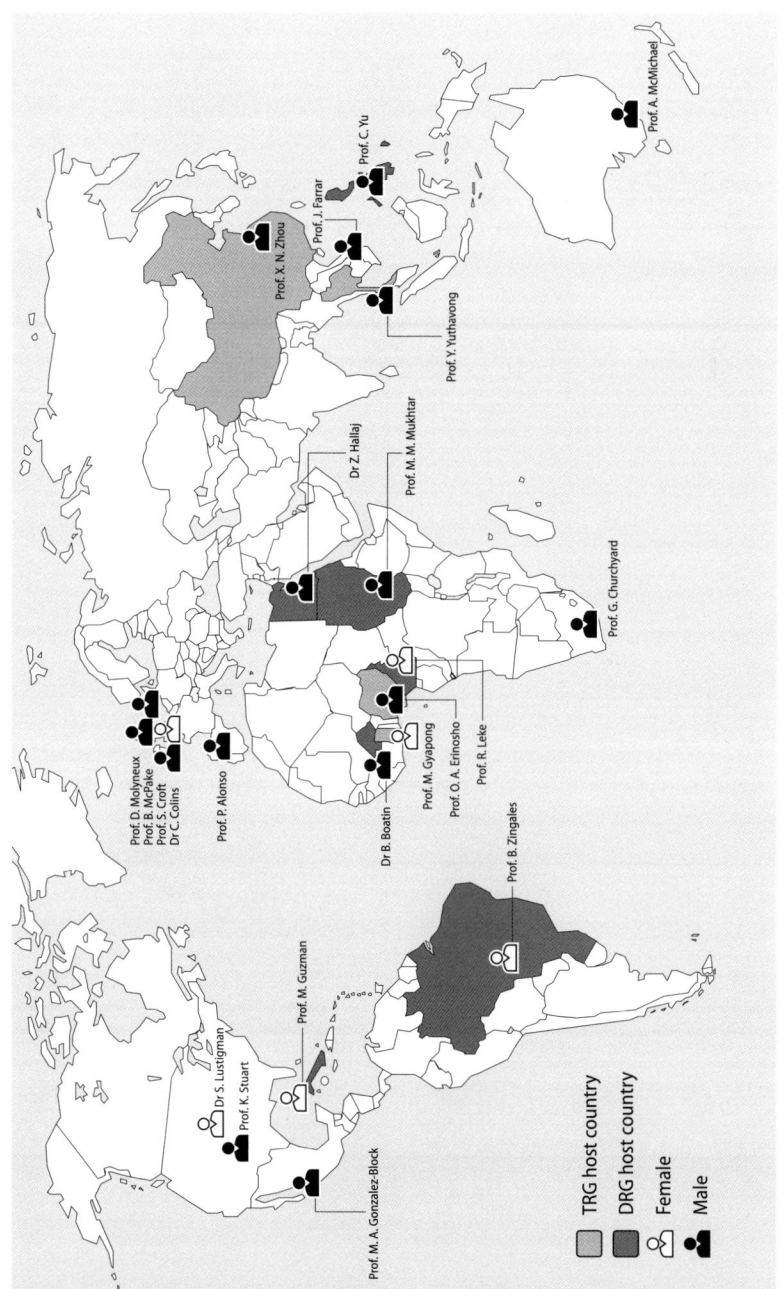